ON THE BRINK OF UTOPIA

 Ideas Series

Edited by David Weinberger

The Ideas Series explores the latest ideas about how technology is affecting culture, business, science, and everyday life. Written for general readers by leading technology thinkers and makers, books in this series advance provocative hypotheses about the meaning of new technologies for contemporary society.

The Ideas Series is published with the generous support of the MIT Libraries.

Hacking Life: Systematized Living and Its Discontents,
Joseph M. Reagle Jr.
The Smart Enough City: Putting Technology in Its Place to Reclaim Our Urban Future, Ben Green
Sharenthood: Why We Should Think before We Post about Our Kids,
Leah A. Plunkett
Data Feminism, Catherine D'Ignazio and Lauren Klein
Artificial Communication: How Algorithms Produce Social Intelligence,
Elena Esposito
The Digital Closet: How the Internet Became Straight, Alexander Monea
On the Brink of Utopia: Reinventing Innovation to Solve the World's Largest Problems, Thomas Ramge and Rafael Laguna de la Vera

Thomas Ramge
Rafael Laguna de la Vera

ON THE BRINK OF UTOPIA
Reinventing Innovation
to Solve the World's
Largest Problems

foreword by
Nobel Laureate Stefan Hell

The MIT Press
Cambridge, Massachusetts
London, England

The MIT Press would like to thank the anonymous peer reviewers who
provided comments on drafts of this book. The generous work of
academic experts is essential for establishing the authority and
quality of our publications. We acknowledge with gratitude the
contributions of these otherwise uncredited readers.

This book was set in
Monument Grotesk by Swiss duo Kasper-Florio and Dinamo
Times Now by Jan-Henrik Arnold

Printed and bound in the United States of America.

Library of Congress Cataloging-in-Publication Data

Names: Ramge, Thomas, 1971– author. | Laguna de la Vera, Rafael, author.
Title: On the brink of Utopia : reinventing innovation to solve the
world's largest problems / Thomas Ramge and Rafael Laguna de la Vera ;
foreword by Nobel Laureate Stefan Hell.
Description: Cambridge, Massachusetts : The MIT Press, [2023] |
Series: Strong ideas | Includes bibliographical references and index.
Identifiers: LCCN 2022061582 (print) | LCCN 2022061583 (ebook) |
ISBN 9780262546485 (paperback) | ISBN 9780262376266 (epub) |
ISBN 9780262376259 (pdf)
Subjects: LCSH: Technology—Social aspects. | Solutionism. |
Technological innovations.
Classification: LCC T14.5 .R347 2023 (print) | LCC T14.5 (ebook) |
DDC 303.48/3—dc23/eng/20230124
LC record available at https://lccn.loc.gov/2022061582
LC ebook record available at https://lccn.loc.gov/2022061583

10 9 8 7 6 5 4 3 2 1

publication supported by a grant from
The Community Foundation for Greater New Haven
as part of the **Urban Haven Project**

FOREWORD

On the Brink of Utopia is a must-read for anyone with an interest in the future of our planet. This highly readable book is not simply about questions of maintaining or reaching more prosperity and well-being. It is about successfully tackling the world's most pressing and, in part existential, problems. With a pronounced basis in techno-optimism, it asks the following questions: Despite all the challenges we face, what factors would make this approach to problem-solving work? How do we stand a chance? What needs to change so that human inventiveness and creativity can save us?

Thomas Ramge and Rafael Laguna de la Vera go right to the heart of the major shifts needed in our innovation culture at large. With a broad outlook on successful role models, the authors systematically analyze what is holding us back from unleashing humankind's incredibly effective problem-solving skills.

As the authors distinguish at the outset, the object is true innovation. Not the kind of innovation that most people could easily have done without. Not the numerous "innovations" that ultimately get the world into even more trouble. *On the Brink of Utopia* discusses the kinds of urgently needed, pivotal developments that will make a fundamental, positive difference in the trajectory of our world.

In seven chapters, Ramge and Laguna explore the cosmos of breakthrough innovations. You can dive into the individual chapters and find interesting observations and analyses on the nature of fair competition, risk taking, differences in financial return models between state and capital markets, the start-up and venture capital world and funding cycle, and much more. Based on a humanistic perspective and a great deal of realism, the book never loses sight of guiding principles of the broadest consensus, the foremost being the 17 development goals of the United Nations.

While reading or listening, you will inevitably find yourself debating with yourself or your friends. The book's critical observations make you think, for example, on the subject of the generally stronger risk aversion in Europe, and the need for a truly sizeable Future Fund if we are serious about investing in the future. We only have one.

The authors are outspoken about the entrepreneurial state, about fresh policy thinking and the role of targeted investments in innovation in the United States, and about what we need more of: new financial vehicles in public and private partnership in addition to precommercial procurement. There are also numerous important observations on funding and the nature of evaluations that are worth thinking about as ways to move toward a more goal-achieving system. Just as important, the need for a culture shift toward a collective that understands "risk intelligence" (Gerd Gigerenzer) is convincingly argued.

Ramge and Laguna emphasize the all-important insight "No risk, no leap," which arguably keeps us from changing the game in many critical areas. With this comes the requirement to embrace honorable failure in trying. "It's acceptable to fail. What's not acceptable is not knowing if you've failed" (Ken Gabriel). I agree.

The characterization of breakthrough innovators and their personal traits near the middle of the book is an especially enlightening read. Drawing on a few carefully selected examples from history and more recent times, the authors brilliantly reveal the essentials of the human beings that change the course of history with their ingenuity. A great deal of trust in this human ingenuity shines through, and I so strongly share this view.

These innovators could be anyone, or could they? Thinking about all the hurdles along the way and the decisions that humans make, you will wonder how much potential for major advancements is lost day in and day out. The breakthrough idea itself, while it can be as short as a moment of deep insight, is not the whole story. Just having an idea is cheap, they say, if you cannot judge its viability and discover its true potential. Although many ideas are being discussed in many places around the world, breakthrough ideas are rare and oftentimes go unrecognized by the public—if not for certain individuals. The challenge is to identify these individuals, who not only come up with breakthrough ideas but also recognize their transformative nature and do what must be done to realize that transformation. These true innovators choose to turn their ideas into facts.

The issues discussed in *On the Brink of Utopia* are so critically important because we must leave the tempting path of "more shades of the same" in many more ways than one would think. This does not mean to do away with rules or being careless, far from it. In Western democracies, control and feedback are built in such that hardly anything will happen with public funding without the consequences being voiced.

"What if we dreamed bigger?" It may come as a surprise, but in the sciences and technology, dreaming is even more justifiable than in fiction. It's because Nature eventually grounds us to reality. Nature will tell us what is possible and what is not. Therefore, to find solutions to humankind's urgent needs, we have to "dream" big and question Nature every day. It will have the final say anyway.

Stefan Hell
Physicist and 2014 Nobel Laureate in Chemistry
Göttingen, August 2022

PRELUDE

THE GREAT INNOVATION LEAP

WHAT IF ...?

What if we had green energy from wind and solar power, hydro-power, and nuclear fusion so inexpensive that it would hardly be worth billing customers for it? With carbon-free energy for one cent or less per kilowatt hour, poverty, drought, and hunger could be radically reduced worldwide. We could remove vast quantities of carbon dioxide from the atmosphere and halt climate change. We could secure access to clean water for everyone. Consequently, the world would become a much more peaceful place. Fewer people would have to flee their homes.

What if a much better understanding of the blueprint of life would finally help us to find cures for all the major diseases: for cancer and dementia, cardiovascular and autoimmune diseases, mental illness and paralysis, blindness, and, of course, all forms of old and new viruses? In other words, cures for all those diseases and threats to health that we cannot get a grip on despite the enormous research effort undertaken in recent decades. How much suffering would that take from how many people? How many new opportunities in life would arise for all those who suffer from these diseases and for their families and loved ones? And what if, thanks to the molecular-biological revolution, we could actually slow down the cellular aging process so that we can remain healthier as we grow old?

A GENUINELY INNOVATIVE LEAP DOES MORE THAN MAKE OUR LIVES A LITTLE EASIER, IT FUNDA-MENTALLY IMPROVES THEM.

What if we would be able to preserve biodi-versity and strengthen protections for animals by radically reducing the amount of land needed for food production thanks to ultra-intensive agri-culture, vertical farming, and resistant breeds? Imagine that before long, meat indistinguishable from what is currently at the supermarket will no longer come from feedlots but from gigantic petri dishes. And what if we really managed not only the nitrogen cycle but cradle-to-cradle cycles for most, if not all, materials and products, and scarce resources became abundant?

What if artificial intelligence (AI) assistants would help us with our decisions, even the difficult ones, and what if these assistants would truly represent our interests and not those of Google or Amazon? And what if augmented reality tools could help us with difficult tasks in professional and private life?

What if education all over the world benefited from digital innovative breakthroughs? Learning could be as fun as a good video game, with robot teachers and human educators to coach peer learning in small groups. What if this approach to education were even a little addicting?

What if we could fly in shared, autonomous, electrically powered drones and didn't need to build roads anymore? What if there were an abundance of cheap CO_2-neutral fuels for long-haul flights? How about shortening the time for a flight to India or Australia with a quick detour through low Earth orbit? It wouldn't hurt to have a system to divert large asteroids that are heading for Earth. And although at least one of the two authors of this book wouldn't be willing to fly along, what if we had a permanent colony on Mars by 2050 with a thousand or so inhabitants? Would this help the human race to rediscover the spirit of discovery and challenge scientific frontiers in order to solve the largest problems on Earth?

What if we dreamed bigger? What if the heroes in these utopian dreams were scientists and technologists, for whom progress and innovation were two sides of the same coin? This book's answer is that we live on the brink of utopia. With science and technology, we can solve all the challenges just described within the next three decades. By 2050, we can live in a world that we want and need because we have the knowledge, the methods, and the financial means to create it. We are being held back by technological incrementalism, fake innovation, and the lack of an optimistic vision of a greener, healthier, and wealthier future. To cross the brink we must reinvent innovation. We need to think in leaps, not steps. Scientists, governments, and markets must work together to ensure that radical innovations do not perish prematurely in the innovator's valley of death. But first, we must admit that we live in less innovative times than we often think we do.

A genuinely innovative leap docs more than make our lives a little easier, it fundamentally improves living conditions for humankind. Think of the cultivation of einkorn wheat, the first sailboat, or the printing press. Think of the water closet's effect on the spread of disease, how the use of fertilizers made urban living sustainable, or the way computers and the World Wide Web have affected just about everything. Then think ten years ahead about how mRNA, nuclear fusion, and vertical farming might have as big an impact. This book is not about the next TikTok or iPhone. Silicon Valley's and China's platforms may well make our lives more convenient, but are they as "disruptive" as they claim? We knew how to shop before Amazon, and order taxis without Uber. And hasn't political discourse become much more divisive thanks to Twitter?

Conversely, in the last two decades science and technology failed to create the radical advancements we yearn for. We made little progress fighting cancer and basically in stopping dementia. Mental health gets worse, not better, as we enter a pandemic age with few antiviral drugs. We don't have enough green energy, and we lack the technology that could suck enough CO_2 out of the air to restore the climate's balance. According to the UN, more than 800 million people are undernourished today—that's a tenth of the global population. Meanwhile, we don't know how to cope with a growing world population and end the overexploitation of Earth's now rapidly depleting resources. This has to change fast.

This book outlines a new, simple framework for making innovation leaps more likely for radically improved technology that can help solve the true challenges of humankind. In seven steps presented in seven chapters, we rearrange the roles of innovators, citizens, govern-

ONLY INNO-
VATIVE BREAK-
THROUGHS
WILL HELP US
ESCAPE FROM
THE ECO-
NOMIC AND
ECOLOGICAL
PATH DEPEN-
DENCIES
WE'VE BEEN
FOLLOWING
SINCE INDUS-
TRIALIZATION.

ments, and financial markets to foster radical innovations that maximize the well-being of the greatest number of people:

→ 1.

Innovation assessment:

We live in an age of innovation theater. Tech development in the last fifteen years has mainly been incremental. It is high time, and a good time, to aim high. Many basic technologies, from AI over molecular biology to advanced materials, are ready to enable radical innovation. We do not need any more apps, gadgets, platforms, and digital business models that supposedly make our lives easier, but in fact infantilize and control us. What we precisely don't need is the kind of sham innovation for which there is almost unlimited venture capital available worldwide.

→ 2.

Refocus:

Climate change, health, poverty—there is clearly no shortage of big challenges where innovation leaps are dearly needed. Indeed, incrementalism can be counterproductive by deepening path dependencies. Therefore, technology must finally focus on solving real problems for true human needs. For this, we develop a model that we refer to as Maslow's hierarchy of innovation, and we tie this model to the UN's 17 sustainable development goals. Maximizing the happiness of as many people as possible, rather than the profits of Big Tech companies, requires a mission-oriented culture of innovation. In this culture, citizens are involved in identifying the challenges, and governments play a far more active role in helping to make them a reality.

→ 3.

Nerds with a mission:

No one does innovation alone. Yet innovation leaps need great individuals with exceptional character traits: an obsessive interest in a particular field, exceptional grit, and being deeply motivated to make a real difference where it really matters. Meanwhile, these innovation leaders also need a particular openness toward recognizing and accepting important ideas and impetuses from other people and, finally, the ability to transmit their own enthusiasm to others and to build and lead teams without descending into micromanagement. This combination is rare. Societies must offer nerds with a mission and the teams around them the best possible opportunities for tech development, with freedom, recognition, and, of course, funding.

→ 4.

New government involvement:

Radical innovation must always solve a chicken-and-egg-problem. When left alone, markets often fail. Governments must therefore embrace their role as boosters of technology readiness, making use of their purchasing power to leap through the "Valley of Death" of radical innovation in which essential ideas die for lack of a secure return on investment. That doesn't necessarily mean vaccines and quantum computers. A thousand miles of road surfacing that resists cracking during cold winters or 100,000 environmentally friendly apartments for $165 per square foot would represent innovative breakthroughs for society. New government involvement can also mean less involvement through regulation, especially in Europe. And removing red tape from state innovation funding could help radically. DARPA (Defense Advanced Research Projects Agency) and ARPA-E (Advanced Research Projects Agency–Energy) can teach many meaningful lessons to many institutions here in the United States and in the rest of the world.

→ **5.**

Reinventing venture capital:

We have an abundance of venture capital for ventures that are not that risky. Most of this money goes to digital services and platforms that have already been proven to work elsewhere. Venture capitalists who want to make a dent must learn (or relearn) to take risks—and take closer looks at deep-tech investments aligned with serving human-kind in significant and even urgent ways. In the long run, it will pay off for them—and for humanity's future. First movers in VC have already realized that in times of technological paradigm shifts, the biggest risk lies in not taking any risks and instead relying on the linear continuation of present trends.

→ **6.**

Reinventing (open) innovation:

Gifted individuals play a crucial role in making innovation leap, but tech will progress rapidly only if many of these great innovators engage in global collaboration with an open mindset. To make innovation leap, we can draw lessons from open-source software development, open access to data in science, and corporate open innovation. This also includes striking a meaningful balance in patent law. The incentive to invest in innovation needs to be preserved. In return, licensing with so-called fair-use models needs to be simplified in order to accelerate the spread of technology.

7.

Picturing utopia:

Don't be careful what you wish for. Reinventing innovation will not succeed unless we dream bigger dreams. What would a world in the year 2050 look like in which science and technology have solved many of today's real problems? This kind of dreaming cannot be left to nerds with a mission. We need a global conversation about scenarios that we should aim for. This book is an attempt to start this conversation.

On the Brink of Utopia has been written by two self-described rational tech-optimists. We tend to speculate with enthusiasm how science and technology will tackle the world's biggest problems, without just forecasting that in the end everything will work out somehow. We will venture another forecast: Some of you, dear readers, will perceive our vision of the future in this book as too tech-oriented and tech-optimistic. Some will find our optimism even counterproductive, arguing that too much trust in technological solutions might hinder a necessary change in human behavior. We can understand that, at least to an extent. Innovation as a fetish has not always helped to advance progress. We understand that innovations have the burden of proving that they are better than what came before—and not the other way around. And we know that there are no magic bullets for complex human challenges, neither political nor technical. All that is correct. But in view of multiple existential threats to humanity, the tried-and-true methods that we have inherited no longer seem to be viable solutions for the future.

With innovative breakthroughs, we will bring about the necessary changes for the economy and the environment, physical and mental health, equal participation and fairness in a global society.

Pessimism is a waste of time, leaves us in a bad mood, and spoils our lives away. Preventing the worst, preserving the status quo, and avoiding experimentation are not answers to the challenges of our time. That is why, at the end of this book, we will sketch out a future in which innovative leaps make life even better than it is today in all its colorful diversity and with all its opportunities and pleasures. In this scenario for a desirable future, technology will solve real problems like obesity and menstrual pain. Arms manufacturing will no longer be financially viable, while politicians and citizens will conduct debates based on evidence using a super Wikipedia, instead of based on aggression using Twitter. In this future, laws and regulations will have expiration dates. Whatever makes things unnecessarily complicated will automatically disappear at some point. We will also speculate about possible shortcuts to the nearest exoplanet. Admittedly, this extraterrestrial jaunt is intended as a mental exercise. Because before our robotic descendants can undertake interstellar travel, we will have to risk major leaps in progress just within the confines of our own planet. The launch pad for this venture was built by German philosopher Ernst Bloch.

Bloch recognized the *constructive power of the concrete utopia* and the role of desires and longings on the way there.[1] It is not as difficult to imagine a future with an abundance of green energy and highly intelligent resource management, radical medical advancements, first-class digital education for all children all over the world, and poverty being an ugly phenomenon of the past. With the current advancements in science and technology, in climate tech and life sciences, in information technology and new materials, in modeling and space exploration, the leap over the brink to such a future is realistic. If we believe in it. If we believe in the human ability to make innovation leap.

NNOVATION THEATER

DO WE ACTUALLY LIVE IN INNOVATIVE TIMES?

12

CHAPTER 1

THE BEST THING IN THE WORLD

Around 11,500 years ago, a genius living in the Middle East, probably on the southern edge of the Taurus Mountains, came up with an interesting idea. This woman—or man—put a seed in the ground and waited until the plant germinated, grew, and finally bore fruit. Today we don't know how this genius came up with the idea or what observations preceded the experiment. We don't know which plant it was—perhaps a wild grain or a berry bush—or whether the innovator watered or otherwise tended the first cultivated plant in human history. It's unlikely. Thousands of years passed before watering cans and pruning shears appear in the archaeological record. But one thing is certain: the idea of cultivating plants, its implementation and wide-scale adoption, radically changed the world.

The ability to grow crops let the people of the later Stone Age settle down and systematically practice agriculture and animal husbandry. Our ancestors learned to make fire around 700,000 years ago. But it was only the successful experiment with seeds that made the leap possible from nomadic hunting and gathering to farming. The Neolithic Revolution had begun, and the field was ripe for technological and social progress.[1]

What was the most important innovation in human history? Was it the ability to grow plants, or was it the wheel, which at around 6,000 years old is significantly more recent than cultivated plants? Or was it the first calculating machine, the abacus, invented less than 5,000 years ago? Paper money, the steam engine, or pasteurization? Measuring daily time with clocks (invented in the fifteenth century), the printing press, or electricity? The radio, the airplane, or reinforced concrete? Could the assembly line make it into the top ten? Or penicillin, the birth control pill, ARPANET as the forerunner of the internet, or that iconic innovation of our time, the iPhone, brought into the world by a man whose goal was to "make a dent in the universe?" Do writing and democracy also count as innovations, and do they belong on this list?

Ranking the greatest inventions of all time is a fun game to play, especially with children and teenagers. The search for an answer gives us a sense of how knowledge and technology, often in connection with social techniques, have repeatedly pervaded societies, spread regionally, and then permeated globally and with increasing speed, leading to spurts in progress. A list of the greatest innovations of all time could only be scientifically justified, however, if we could define consistent criteria for progress across millennia of technological development and then quantify this progress using meaningful and consistent units of measurement. This would be a difficult, probably even impossible, undertaking. The criteria would have to be normative, so they would themselves be derived from subjective values and preferences. In any case, over a time scale of thousands of years, we lack basic data for establishing a ranking. But a few years ago, the American historian of technology, Leslie Berlin, supported by a group of twelve innovation experts, scientists, entrepreneurs, philosophers, managers, and authors, nevertheless attempted to systematically approximate the fifty most important breakthrough innovations in human history.[2]

Berlin's first step toward an objective ranking was to develop a taxonomy of great innovations: a systematically meaningful subdivision into fields of application in which science and technology led to progress. This resulted in a total of seven categories: innovations that "expand the human intellect," are "integral to the physical and operating infrastructure of the modern world," "extend life," "allow real-time communication," "move people and physical goods," are "organizational breakthroughs," and "enabled the Industrial Revolution." The experts on the panel then submitted

1. THE PRINTING PRESS, 1430S
2. ELECTRICITY, LATE 19TH CENTURY
3. PENICILLIN, 1928
4. SEMICONDUCTOR ELECTRONICS, MID-20TH CENTURY
5. OPTICAL LENSES, 13TH CENTURY
6. PAPER, 2ND CENTURY A.D.
7. INTERNAL COMBUSTION ENGINE, LATE 19TH CENTURY
8. VACCINATION, 1796
9. THE INTERNET, 1960S
10. STEAM ENGINE, 1712
11. NITROGEN FIXATION, 1918
12. SANITATION SYSTEMS, MID-19TH CENTURY
13. REFRIGERATION, 1850S
14. GUNPOWDER, 10TH CENTURY
15. AIRPLANE, 1903
16. PERSONAL COMPUTER, 1970S
17. COMPASS, 12TH CENTURY
18. AUTOMOBILE, LATE 19TH CENTURY
19. INDUSTRIAL STEELMAKING, 1850S
20. BIRTH CONTROL PILL, 1960
21. NUCLEAR FISSION, 1939
22. GREEN REVOLUTION, MID-20TH CENTURY
23. SEXTANT, 1757
24. TELEPHONE, 1876
25. ALPHABETIZATION, FIRST MILLENNIUM B.C.
26. TELEGRAPH, 1837
27. MECHANIZED CLOCK, 15TH CENTURY
28. RADIO, 1906
29. PHOTOGRAPHY, EARLY 19TH CENTURY
30. THE MOLDBOARD PLOW, 18TH CENTURY
31. ARCHIMEDES' SCREW, 3RD CENTURY B.C.
32. COTTON GIN, 1793
33. PASTEURIZATION, 1863
34. GREGORIAN CALENDAR, 1582
35. OIL REFINING, MID-19TH CENTURY
36. STEAM TURBINE, 1884
37. CEMENT, 1ST MILLENNIUM B.C.
38. SCIENTIFIC PLANT BREEDING, 1920S
39. OIL DRILLING, 1859
40. SAILBOAT, 4TH MILLENNIUM B.C.
41. ROCKETRY, 1926
42. PAPER MONEY, 11TH CENTURY
43. ABACUS, 3RD MILLENNIUM B.C.
44. AIR CONDITIONING, 1902
45. TELEVISION, EARLY 20TH CENTURY
46. ANESTHESIA, 1846
47. NAIL, 2ND MILLENNIUM B.C.
48. LEVER, 3RD MILLENNIUM B.C.
49. ASSEMBLY LINE, 1913
50. COMBINE HARVESTER, 1930S

their top breakthrough innovations according to these categories, and a vote was taken.

It was also clear to the panel of experts that the list couldn't claim to be complete or objective. But, presented as pictures on a timeline, it provides a good overview of when science and technology have advanced human progress and improved our lives, and in which areas. It clearly shows the acceleration of technological progress from the middle of the nineteenth century up to the 1930s, when the great discoveries were rapidly made and then turned into the technological solutions and products that so strongly shape our lives, medicine, the economy, and the organization of society to this day.

A look at this diagram leaves even techno-optimists like the authors a little thoughtful and deflated in light of the accomplishments in innovation of past generations. The diagram raises the question: Have we perhaps long since passed "peak innovation," the pinnacle of technological and creative ingenuity, and are innovative leaps becoming increasingly rare? It might be that the low-hanging fruit of progress has long since been harvested—like that of the Neolithic genius who planted a seed in the ground and waited.

James Fallows, "The 50 Greatest Breakthroughs Since the Wheel," *The Atlantic*, November 2013

HYPERSCALING AND THE GREAT STAGNATION

Are we living in especially innovative times? Or do we suffer from vanity of the present, overstating current innovation because who wants to live in boring times? Is technological progress in fact only taking baby steps forward despite all the hullabaloo about innovation? The vanity of the present would in this case be the result of a frenzied technological standstill.

Thought experiment: A few friendly alien innovation researchers come to our planet. They conduct numerous in-depth interviews with their local colleagues, experts, and economic actors in the broad field of innovation and technology about the overall situation of innovation on planet Earth. Presumably, the overall diagnosis of these friendly extra-terrestrial observers of humanity would be deep disagreement and, with many individuals, deep inner turmoil. Because let's imagine . . .

The extraterrestrial innovation researchers first decide to bask in the energy of start-up conferences in San Francisco, Lisbon, and Shenzhen. There they would listen to start-up founders' TED Talk–ready presentations full of narratives about disruption and hyperscaling, which will not only generate obscene profits for the start-ups but also make the world an all-around better place. The aliens would also follow powerful American and Chinese venture capitalists on Twitter and LinkedIn, as well as Ray Kurzweil (the Singularity!), Brian Roemmele (consider!), and the TechCrunch and Techmeme account. The visitors would attend the opening ceremonies of innovation labs, innovation hubs, and the business incubators of major corporations. They would listen to AI researchers, blockchain entrepreneurs, and design-thinking consultants and study "roadmaps for digital transformation." At a dinner held at an outrageously expensive restaurant in Silicon Valley, the extraterrestrials would try lab-grown meat that really tastes like beef. They would share a table with people claiming they will soon be able to stop the cellular aging process and even defeat death. Perhaps the visitors would also drink coffee of a conventional sort in Brussels, Washington, and Beijing with the politicians and bureaucrats who make and administer research policy and study funding plans with numbers ending in lots of zeros, clearly supported by anticipated timelines for quantum computers and nuclear fusion.

On the way to the hotel, the guests would read on TechCrunch that a certain Elon Musk was one of the richest people on Earth. They would learn that this Musk was not only sending the internal combustion engine off to the technology museum and building a self-driving car but was also running a very successful rocket company on the side and would like to found a colony on Mars soon with people who had been upgraded to cyborgs using his "Neuralink." At least by the time they learned that another super-rich person (depending on current stock prices), Jeff Bezos, who in the last two decades has used recommendation algorithms to revolutionize shopping and cloud computing to revolutionize logistics and computer applications, is now exiting his core business so he can also fly into space on his own rocket ships—at that point at the latest, the impression would have taken hold among these friendly innovation researchers from a distant exoplanet that Earth people really are extremely innovative. Suitably impressed, they could depart for home—unless they first met smart but skeptical minds, such as Tyler Cowen, Robert J. Gordon, and Lee Vinsel.

Perhaps the three economists, with expertise in the history and sociology of technology, would press copies of their books *The Great Stagnation*, *The Rise and Fall of American Growth*, and *The Innovation Delusion* into the hands (or other appendages as appropriate) of these guests from outer space.[3] After reading the books, the visiting aliens would be deeply unsettled because economic, historical, and cultural indicators also pointed to the opposite conclusion: the human race is currently experiencing an innovative dry spell, and no one knows if it will ever end.

The most important economic argument that we're currently living in a phase of diminished innovation is provided by a historical comparison of increases in productivity. With numeric data and examples, Cowen and Gordon demonstrate that the major increases in prosperity between 1870 and 1970 were driven by increases in productivity due to technology. The great inventions and their rapid spread during the "golden age of innovation"—electricity and the telephone, railroads and the automobile, plastic and concrete, indoor plumbing and antibiotics, television and refrigerators—led to a one-time boost through technology, not to be repeated for the foreseeable future in the history of innovation. The two economists see the uniqueness of the event in the transition, along a path paved by technology, from an agrarian economy to an industrial and service economy. The leap so many people talk about today—from industrialization to a knowledge economy founded above all on information technology—seems rather insignificant in comparison. In addition, this transition is evidently fluid. You could even call it sluggish, as the so-called knowledge economy still has strong industrial characteristics.

Neither Tyler Cowen nor Robert Gordon consider themselves enemies of technological progress. They have nothing against social breakthroughs with the help of technological solutions, and they also don't belong to those who preach renunciation of prosperity and pleasures. They merely estimate the degree of innovation and the impact of technological developments since the internet achieved its World Wide Web–era breakthrough to be much lower than the Web's euphoric developers, financiers, salespeople, and users often do. Viewed rationally, according to the two economists, online shopping, social media, car-sharing apps, video conferencing, and smartphones are only incrementally better applications in comparison to their predecessors. Correspondingly, these recent tech innovations don't generate significant gains in prosperity as do real technological breakthroughs, such as the bicycle, which at one point extended many people's radius of movement in everyday life from 3 miles as a pedestrian to 12–18 miles. In terms of economic benefit, the bicycle is even today likely to be one of the most underrated innovative leaps—and, as the current bicycle boom shows, also one of the most long-lasting. Steve Jobs knew that too when he referred to the computer as the "bicycle for the mind." [4]

With scientific detail, the skeptics among progress researchers base their less rosy assessment above all on the historical time series of so-called total factor productivity. This term describes the part of productivity growth that economists can't assign to changes in the classic production factors of labor, land, and capital. It is plausible that the "unexplained remainder" of rises and falls in productivity is strongly related to the use of new technologies. If you follow these basic assumptions, in light of this time series, it also seems plausible that the innovation gains that were comparatively easy to obtain were achieved with the major innovations of the nineteenth and twentieth centuries. A similar development could only be expected again if a major socioeconomic change becomes possible through technology, such as at the transition from a hunter-gatherer society to agriculture and animal husbandry, or from the early modern agricultural economy to industrialization.

To demonstrate that the digital revolution has a significantly lower level of invention than is often claimed, Robert Gordon could bring the extraterrestrial innovation researchers along to a lecture hall at Northwestern University in Evanston, Illinois. At eighty years old, he's still teaching. He regularly conducts an interesting thought experiment with his digital-enthusiast Generation Z students: the toilet test. [5]

TOTAL FACTOR PRODUCTIVITY (USA)
average annual increase per decade

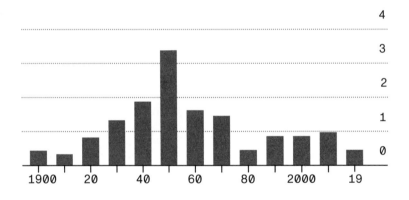

Source: Robert J. Gordon;
Federal Reserve Bank of San Francisco

The young people are asked to decide between two options:

Option 1: You have a 2010-era laptop or personal computer running Windows 7 and the usual programs of the time. You have access to the internet according to the standards of 2010. You also have a toilet inside your apartment.

Option 2: You have the latest MacBook with an internet connection and a smartphone of the latest generation with access to all apps, social media, and scooter sharing. There is an outhouse in the backyard.

Although the comparison is imperfect, as we'll see when we take a look at Maslow's hierarchy of needs in the next chapter, in the opinion of most students, the usefulness of a toilet within your own four walls beats the informational added value of TikTok. Lee Vinsel, a much younger economist and sociologist of technology, offers his readers a similar thought exercise in *The Innovation Delusion*: with the book in hand—and most of the time, it's a printed book and not an eBook on a digital device—they're supposed to look around the room.[6] Then they're supposed to count how many technologies around them are older than fifty years, that is, were invented before 1970—and which technological objects are younger than Generation X, to which Vinsel himself belongs. Take a quick look around of your own. And?

The title of Vinsel's book (coauthored by the historian of technology Andrew Russell) already reveals what he thinks of the narrative of

ON THE STAGE, THE PEOPLE PLAY-ACTING INNOVATION HAVE PERFECTLY MASTERED INNOVATION TALK.

22

CHAPTER 1

rapid progress in the third millennium, fueled by the "move fast and break things" mantra developed in Silicon Valley and exported worldwide. The term "delusion" is interchangeable with "deception" or "illusion," if you prefer. After taking the pulse of application-oriented research, Big Tech corporations, and start-ups, the authors' diagnosis is an almost pathological mixture of hubris and denial of reality; the more obvious it becomes that major leaps in innovation are failing to occur, the more the technology marketing machine turns up the volume. In a nutshell, their central thesis is that "we do not live in innovative times, but in the age of innovation theater."

On the stage, the people play-acting innovation have perfectly mastered innovation talk. For Vinsel and Russell, however, the inflation in the number of innovators with speaking roles, presented by "Chief Innovation Officers," "Innovation Evangelists," and "Heads of Design Thinking," is not only tedious, but also hazardous on at least two levels. On the one hand, when it comes to creating harmful products and companies, from insecure software to the Wirecard fiasco, innovation theater is a repeat offender. But on a deeper level, in Vinsel and Russell's view, the endless praise for the new and disruptive solution built on a new and disruptive technology undermines the foundation on which our prosperity and a good life for the majority of the world's population rests: the maintenance of the existing systems and infrastructures that produced the golden age of innovation. Unfortunately, maintenance is not a winning topic for grandiose TED Talks.

The pseudo-innovators talk about hyperloops and artificial superintelligence, while in the here and now, roads and railways are becoming dilapidated, and bridges occasionally collapse. The iconic disaster of delayed maintenance is the Morandi Bridge in Genoa: opened in the 1970s as a marvel of the latest bridge technology, allowed to decay over five decades, ultimately becoming a death trap for forty-three people in 2018. Presumably, Vinsel and Russell would have to assure the extraterrestrial innovation researchers that the bridge really did collapse, even though the human race is so advanced, and the Romans built giant viaducts over two thousand years ago that, like the Pont du Gard in southern France, still cross the landscape with amazing stability. This is one of the reasons why the authors of the current book intend to create an international communications platform called The Maintainers. Perhaps the extraterrestrials would find evidence for doubts about innovation there as well. If they stumbled upon news articles on a certain Elizabeth Holmes, who set out to revolutionize medicine with highly automated blood testing with her biotech start-up Theranos and was listed among the 100 most influential people in the world but who ended up in jail for criminal fraud, the friendly alien innovation researchers might be totally confused about the state of mind of the human race.[7]

THE FUTURE REMAINS UNPREDICTABLE

Are the skepticism toward and criticism of today's innovators completely exaggerated? And isn't it normal and even helpful for new technologies to be a bit exaggerated, or for their inventors to promise the Moon? After all, that's the only way for them to get attention and start-up capital and thus the chance to be tested and survive the long march through innovation's valley of death to ultimately create economic value as a useful part of our lives.

At the moment, no one can accuse the human race of being stingy or putting too few bright minds to work on the hard problems of innovation. The budgets for research and development are growing just as rapidly as the number of innovators. This isn't just the case in China—although the trend is particularly strong there—but also in the United States, in Europe, and in the innovative economies of Asia.

Or has the current generation of researchers and developers in fact lost the ability to make innovative breakthroughs? Not because they're less capable than previous generations but because it has become much more difficult to make the next leaps in innovation successfully? A 2020 study with the title "Are Ideas Getting Harder to Find?" led by Stanford economist Nick Bloom points in this direction.[8] Using data from the chip and biotech industry, Bloom and his colleagues demonstrate that yes, it is becoming more difficult, and significantly so. In order to achieve the same gains (outputs) from innovation processes, governmental and private organizations have to constantly invest more money and more working hours from researchers (inputs).

The two of us would also like to have a conclusive and reliable answer to the question of whether we're living in innovative times and whether there really are a number of developments in the pipeline with high chances of success and major benefits, or whether technology optimists (like us) are just claiming that. The honest answer to this question is simply that we don't know, nor can we know, the answer; whether an idea will become an innovative breakthrough of great benefit and little harm can only be seen in retrospect. In academic terms: the evaluation of an innovation's benefit occurs ex post, not ex ante. This is necessarily the case because the future remains unpredictable, even in times of analyzing data on a massive scale and using learning algorithms to improve prediction. Then again, we find this reassuring. Otherwise, life would be predetermined and boring, and we wouldn't be able to choose or shape our paths as individuals or societies.

"A prognosis about the usefulness of technology only says something about us today. About our current level of knowledge. But little or nothing about the future."[9] This statement is from Armin Grunwald,

TOTAL GLOBAL SPENDING ON RESEARCH AN DEVELOPEMENT (R&D)
FROM 1996 TO 2022 IN BILLION U.S. DOLLARS (PURCHASING POWER PARITY)

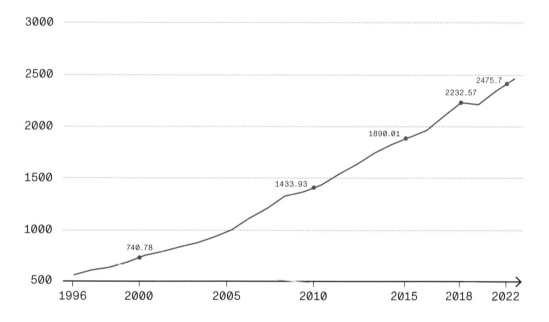

Sources: UNESCO Institute for Statistics; R&D Magazine, © Statista 2022

a German philosopher who studies technology and its impact. His statement is fundamentally valid for all forms of technological progress but especially for radical innovations. Nothing truly new—and certainly nothing radically new—can be calculated from the data of the past and present. First, it would presuppose that nothing unexpected will happen, even though producing the unexpected belongs to the essence of technological change. Second, any innovative leap would already exist, at least in theory, and the technology for it would only be a question of implementation. This in turn means that we can think of the technological future only in scenarios with varying plausibility. If we do this in relation to the capacity for innovation in the future with a medium time horizon, the result is this admittedly coarse-grained model:

Our capacity for innovation in the next ten years could

    a) decrease
    b) stay the same
    c) increase

If we then take a closer look at data from the recent past and the present, we find many indications that scenario (c) is at least plausible and is probably the most likely scenario. We find these indications on five levels:

 1.

Cooperation of the Global Science Community
The COVID-19 pandemic was the most serious (acute) challenge faced by humanity in recent history. When it came to fighting the virus, many governments, health officials, social groups,

and individuals didn't exactly cover themselves with glory. But the biotech community did prove that the current generation of researchers is definitely still capable of innovative leaps when it matters, and in two ways. Several vaccines were developed in China, Russia, the UK, Germany, and the United States in a fraction of the time it previously required. That speed saved millions of lives. Second, in the face of the deadly threat and under the leadership of the companies BioNTech and Moderna, messenger RNA technology achieved a breakthrough. It is evidently a platform for many other potential medical and biotechnical breakthroughs. This triple success gave new self-confidence and new energy to the global research and development community. A new spirit of cooperation is also invigorating the community, because ex post it has become clear that ultra-rapid vaccine development was possible only because scientists around the world shared their knowledge with others with unusual openness. The basis, objective, and overriding characteristic of innovation used to be dominating others based on knowledge. Those days are gone. As we have known since the Enlightenment at the latest, a culture of openness and cooperation provides the best foundation for accelerating innovation. The pandemic marks a turning point that offers abundant evidence for optimism and will result in a regulatory softening on intellectual property rights, which previously have been too inflexible.

 2.

More Private and Public Investment

Innovation has always been expensive, and unfortunately, it's getting more and more expensive. But all the important stakeholders in the grand game of innovation now accept that, and they're raising their stakes. Just by itself, Amazon invested more than 56 billion dollars in "technology development" in 2021 (including digital content). The amount of venture capital available globally sets new records each year, and even in Europe things are slowly getting better. But it's not just Big Tech companies and tech investors who are throwing money at R&D.

The situation with public institutions looks to be even more encouraging. In chapters 4 and 5, we'll go into more detail on the newly discovered role of governments in promoting innovation, but here are a few numbers for now. From 2010 onward, inflation-adjusted research and development funding in the Organization for Economic Cooperation and Development fell continuously, in part due to budget cuts and austerity measures after the financial crisis. But since 2018, government research investments have been reliably increasing again each year, not only in absolute numbers but also as a percentage of gross domestic product (GDP), measured as R&D intensity.

France has announced that it will increase its research budget by 30 percent until 2030.[10] In Germany, long-term plans look similar.[11] As for the European Union, the Horizon Europe Program, passed in 2021, adds an additional 95.5 billion euros in funding for research and innovation until 2027, and is the world's biggest multinational research program to date.[12] US president Joe Biden with the support of Republicans in the Senate has put an end to Donald Trump's antiscience policies. With its 2023 budget the US administration has increased public R&D spending by about 25 percent compared to 2021 totaling more than 200 billion US dollars.[13] On top of that, the CHIPS and Science Act, together with the Inflation Reduction Act, both passed in summer 2022, can be expected to boost research and research spending in the United States in the coming years.[14]

The numbers from China are similarly impressive. In 2022, R&D spending in China has reached yet another high at around 456 billion US dollars, 2.55 percent of GDP and an increase of about 10 percent from the previous year, and the central government plans to increase R&D spending by 7 percent annually through 2025.[15] Globally the R&D intensity relative to GDP has seen a long stagnation since the 1990s, but since 2014 it shows a significant increase from around 2 percent in 2013 to 2.6 percent in 2020.[16] Of course, a lot of input doesn't necessarily mean there will be any output. But financial opportunities and incentives are a necessary prerequisite. And they're increasing.

→ 3.

Growing Competition

We believe in cooperation and openness in the innovation process and, at the same time, in tough and fair competition in a "Darwinian Sea of innovation," a term coined by Lewis Branscomb, long-time IBM chief scientist, head of the National Bureau of Standards and later Harvard professor.[17] Increasing competition is currently invigorating the innovation business in two ways. The high level of public investment by the Biden administration may also have something to do with a fundamental openness to facts and progress through science, in contrast to the previous administration. More important, however, is the increasingly fierce rivalry with China's political and economic system. The continued economic rise of China is closely linked to its emergence as a nation of science and technology that no longer merely copies products and manufactures them more cheaply but is catching up technologically, and in some fields, such as quantum technology, even overtaking its competitors. A reliable indicator for this is the number of high-impact research papers: scientific studies that directly contribute to technical and economic development. Here, too, Chinese scientists are now ahead in many areas.[18]

From a political point of view, the scientific and economic rise of China may seem threatening to us in the West. If it drives us to invest more time (of scientists and developers) and money (both private and public) in technological progress, it will have a positive effect on innovation in the current decade and the following one. The benefits of it will also be reaped worldwide. As is well known, innovation doesn't stop at the borders between political systems, and it isn't a zero-sum game, despite what discussions of technological hegemony often suggest. But even within the tech hemispheres, there's a lot going on in terms of increasing competition. More precisely, in terms of restoring fairer competition.

 # 4.

Revival of Antitrust Regulation

Oligopolies and monopolies are famously rewarding for those who own or run them. Everyone else pays the price. On a second level related to competition and progress, it's encouraging that more and more politicians, citizens, and consumers in the United States and Europe are recognizing and understanding that market dominance by a few Big Tech companies, even extending as far as monopoly positions, slows down progress. In the first and second decade of this century, Apple, Google, Facebook, and other tech giants managed to convince us that they're the drivers of digital progress. At first glance, that seems impossible to deny, and in some respects not even at second glance. But at the same time, we've been blinded by their gigantic R&D budgets, astronomical stock values, and convenient applications that permeate our lives. We have too rarely stopped to ask who digital innovations could have benefited, and by how much, if Silicon Valley hadn't monopolized the data, systematically bought up innovative start-ups at inflated prices, or terminated them with dubious methods in the so-called kill zone.[19] We see increasing market concentration not only with digital platforms but also in traditionally innovative industries, such as chemicals, pharmaceuticals, transportation, and financial services. Advocates of the platform economy, such as Peter Thiel, don't hide the fact that there's nothing more interesting to them than building monopolies.[20]

Monopolies are an act of theft committed against progress, and the world pays twice, through excessively high prices and a slowed pace of innovation. Because more and more people are coming to understand this, the reputation of Big Tech companies that were so celebrated in the '00s has sunk to the level of Big Banking and Big Oil. And not only that, the "techlash" of popular opinion is being followed by a regulatory response to monopolies. In America, Europe, and some of the major emerging economies like India, we're currently seeing a revival of traditional principles of antitrust law. The US Department of Justice, the Federal Trade Commission, and the EU will see to increased competition in the next decade. Lawsuits in the United States against Apple and Google, politically supported by the energetic chair of the Federal Trade Commission Lina Khan, Europe's new Digital Markets Act and Digital Services Act, pushed by the commissioner for competition and executive vice president Margrethe Vestager, are only harbingers of a renaissance of governmental initiative to finally restore fair competition to the field of digital innovation and other innovative industries.[21] Even China's economic bureaucracy is making forays toward reining in the dominance of the digital superstar companies, even though they have long been considered the poster children of China's rise to power. These are all positive indications that in retrospect, the 2020s will go down in history as a decade of technological advancement.

5.

Rise of Public Innovation Agencies

Governments are taking stronger action against monopolies through regulation. They are also becoming more agile in actively driving major innovations. This may not yet be happening with sufficient speed and consistency, but they have at least recognized the problem: since the postwar years, administrative processes have become more and more rigid and formalized to avoid errors from the outset. In order to act more agilely, a number of countries have founded agencies for breakthrough innovations based on the model of the American DARPA. In 2019, Japan launched its government program Moonshots with an initial budget of a bit less than 1 billion dollars for the first five years. Germany with SPRIND and the UK with ARIA followed suit with comparable initial budgets. In April 2022, Canada joined the club, allocating 780 million US dollars in public money designed to boost Canada's private sector's poor track record in innovation.[22] Canada's agency is therefore not modeled on DARPA's high-risk, high-gain approach for innovation leaps but will try to find inspiration from two other state innovation agencies that have proven how much impact government stimulation can have for national innovation ecosystems, the Israel Innovation Authority and the Finnish Funding Agency for Technology and Innovation.[23]

Hence the country that invented state innovation agencies, the United States, currently has the greatest innovation ambitions with them. The Biden administration has founded ARPA-Health for fundamental medical innovation with an initial budget of 1 billion dollars, which is supposed to grow to 2.5 billion dollars in 2024. The United States has also created ARPA-Climate to help climate technologies achieve a breakthrough. ARPA-Climate is now in competition with ARPA-Energy, which has been impressively successful in helping green energy innovations get off the ground since 2009. What does this "rise of ARPA-everything" mean, as Nature wondered, not without smugness?[24] We will go into this in more detail in chapter 4, but at this point it can be said that competition among innovation agencies is good news for the world.

THE NEXT BIG WAVE

In 1926, the Soviet economist Nikolai Kondratiev published an article with the inconspicuous title "The Long Waves in Economic Life."[25] However, the content was as spectacular as the roller-coaster ride the world economy had taken after World War I, with hyperinflation in many countries and a surprising, rapid resurge.

Kondratiev had studied the economic booms and busts in the United States, England, Germany, and France since the beginning of the industrial revolution in the mid-eighteenth century and discovered an interesting pattern. The short, frenetic business cycles, including perennial over- and undershooting of supply and demand, are overlaid by "long waves" with a time horizon of forty to sixty years. In Kondratiev's model, the starting point of the long waves of economic resurgence are fundamental technological paradigm shifts, such as mechanical looms and the steam engine, steel production and railways, electrification and breakthroughs in chemistry. When a basic technology such as this becomes available, national economies, organizations, and individuals massively invest in the new technology and thus create an "innovation-induced upsurge." Once the innovation has established itself across the board, its economic marginal utility will be zero or close to it. Investments then continually decline, and a prolonged downturn follows. In this downturn phase, societies in general and innovators in particular are encouraged or challenged again to put increased effort into the next technological paradigm shift and thus to ensure the next long upsurge in productivity.

For the British economist and Nobel laureate John Maynard Keynes and his disciples, Kondratiev was always an important point of reference and a source of inspiration, although they often went without mentioning him explicitly. Of all people, Keynes's primary intellectual opponent, the Austrian American economist Joseph Schumpeter, continued to make observations and develop Kondratiev's thoughts about long economic waves during his time at Harvard, especially in his 1939 book *Business Cycles*.[26] Schumpeter particularly emphasized the role of basic innovations that create new industries in his famous process of "creative destruction," using Kondratiev's wave pattern as an important source of inspiration. The Russian didn't live to see his work elevated by the "greatest economist of his time," as Schumpeter liked to call himself. In 1938, Kondratiev, the self-taught son of simple farmers, was sentenced to death in Stalin's Great Purge. His model survived—and experienced its own ups and downs.

Mainstream economists have criticized "Kondratiev waves," pointing out that short-term and long-term waves cannot be clearly separated empirically, and that the period of forty to sixty years thus seems to be chosen arbitrarily. Among Neo-Schumpeterian

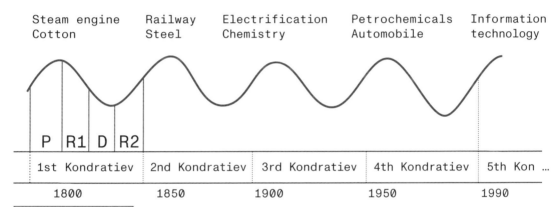

| Steam engine | Railway | Electrification | Petrochemicals | Information |
| Cotton | Steel | Chemistry | Automobile | technology |

P | R1 | D | R2

| 1st Kondratiev | 2nd Kondratiev | 3rd Kondratiev | 4th Kondratiev | 5th Kon … |

1800   1850   1900   1950   1990

P: Prosperity
R1: Recession
D: Depression
R2: Recovery

economists, especially at the University of Sussex, however, the model has seen a renaissance since the late twentieth century.[27] They hope that a sixth Kondratiev wave will soon provide a new, long-term boost in productivity. These scholars place the beginning of the fifth wave around the year 1970, when computers and information technology had reached a tipping point for mass use. Accordingly, the next wave would be due to begin around now.

As with any model, the Kondratiev wave has limitations for describing reality. And our view remains that innovative leaps cannot be predicted because that would contradict their disruptive nature. But the model seems correct and helpful in its systematic view of the connection between the degree of maturity and the spread of new technologies, and the impetus that people regularly need and the trends they establish in order to grapple with problems in the future better than in the past. This is how the world of technology advances cyclically.

EVERYTHING CAN GET BETTER

In the last fifteen years, the momentum of technological progress has been weaker than the euphoric technology utopians have claimed. At the same time, however, a number of basic technologies have developed a level of maturity and proliferation that makes a new wave of innovation at least likely. In some areas, this can already be clearly demonstrated. In addition to the success of mRNA platforms mentioned earlier, other areas include synthetic biology as a whole, with its potential applications in chemistry, medicine, agriculture, and the development of new microchips. The spread of cloud computing, the rapid improvement in machine learning applications, and advances in robotics and nanotechnology are no longer affecting only business plans but also companies' balance sheets. The Internet of Things, conceived by the often overlooked computer visionary Mark Weiser under the catchphrase "ubiquitous computing" more than thirty years ago, has, after considerable delay, finally reached the productive phase.[28]

Energy storage is making rapid progress. Energy efficiency technologies and renewable energy have long since reached the phase of rapid wide-scale adoption in combination with the self-reinforcing effects of broad proliferation. Hydrogen clearly has the potential to do so as well. NextEra, a provider of green energy, has risen to become one of the most valuable energy companies in the world.[29] Deep-sea mining, if it is approached sustainably, and also the new space boom offer further opportunities to turn the '20s of this century into a decade of technological progress, like the 1920s a century ago—this time, we can only hope, without the global geopolitical catastrophes.

Tyler Cowen, Robert Gordon, and Lee Vinsel's skeptical objections to techno-hype and empty talk of innovation should be read as a warning—but even more as motivation. We have to learn to differentiate between real innovation and innovation theater, to promote real innovations and let the innovation big talkers just keep talking. The hypothesis that the period from 1870 to 1930 was a historically unique phase for technological innovation as society transitioned from an agricultural to an industrial economy is, from a scientific point of view, only anecdotal evidence. N = 1. That doesn't seem terribly reliable. Nor would we be the first generation of people to believe that the best years are already behind us. So far in history, this perception has always been wrong, as usual with predictions made by foretellers of doom, who are surprisingly still in business. By and large, things have ultimately gotten better and better through science and technology. Why should it be any different this time?

THE
IS ACT
IS GETT
B

WORLD
JALLY
ING
ETTER

In the science fiction trilogy *The Three Body Problem* by the Chinese author Liu Cixin, the Trisolaran species is on its way to Earth.[30] Unlike the species of extraterrestrial innovation researchers in our little thought experiment at the beginning of this chapter, the Trisolarans are hostile to human beings. They intend to occupy the Earth and eradicate human beings in the process, with the exception of a few subservient vassals. The aspiring colonists come from a planet around 4.2 light-years away. Their fleet is traveling toward Earth at a tenth of the speed of light, so it will take them around 450 years to arrive. The fiendish Trisolarans in their spaceships know that they are technologically far superior to the humans of Earth—at least so far. But they have a well-founded fear that humanity could catch up technologically in the four centuries before their planned hostile takeover and perhaps successfully repel the attack. A fifth column of Trisolarans on Earth therefore use an eleven-dimensional supercomputer to paralyze all of humanity's fundamental innovation. They freeze knowledge and technology in their current state without people noticing it at first.

Epistemologically speaking, we can't rule out with 100 percent certainty that this isn't our situation right now. If it were, we wouldn't know. But we think Cixin's scenario is brilliant science fiction that has no relevance to our current condition. With a probability bordering on certainty, there is no invisible limit to human innovation, neither inherently nor externally induced. The speed at which we drive progress forward has always been, is now, and will always remain a question of ambition and resources. We don't know how much input we'll have to invest for which output in the future. But we have never known that. In times of abundant capital, we can afford it. In the meantime, we should think more about meaningful fields of application and the goals of technology than has been the case in the illustrious past phases of technological innovation.

In the next chapter, we will look for answers to the question, How do we steer innovation in a direction so it will benefit as many people as much as possible?

MASLOW'S HIERARCHY OF INNOVATION

WHAT KIND OF INNOVATION DO WE NEED?

LOVELY ALGAE

During an Arctic expedition around ten years ago, the Spanish marine biologist Mar Fernández-Mendéz experienced a moment of sudden enlightenment. From the railing of her research ship, she could see emaciated polar bears without an ice floe to climb up onto. It was like a scene from a BBC documentary on climate change was taking place in front of her eyes, lacking only the commentary of Sir David Attenborough. The climate change data that Fernández-Mendéz and her colleagues collected on the voyage provided the hard evidence to accompany the scene. Somewhere out in the melting ice pack, she decided, "Researching the causes of climate change as a scientist isn't enough for me. I have to do something to counteract its effects directly—using the tools I bring to the table as a scientist." Today, Fernández-Mendéz is cofounder and chief scientist at Seafields Solutions, a rapidly growing company headquartered in London that aims to extract gigatons of carbon dioxide from the Earth's atmosphere with offshore algae farms.

In many of the world's coastal regions, algae are considered an environmental nuisance, especially where rising water temperatures and CO_2 levels are accelerating their growth. But Mar Fernández-Mendéz talks about these weeds of the sea like flower breeders talk about orchids. "Sargassum is especially beautiful. It's golden, with countless small, semi-transparent beads that make the algae float so gracefully in the water." But the aesthetics of this fast-growing aquatic plant with low nutrient requirements won't stop the Spanish scientist and her colleagues from sinking them en masse to the sea floor. And the more, the better. Their plan goes like this: By 2025 at the latest, there will be huge barriers floating on the high seas. Inside them, Seafields will ensure the ideal breeding conditions for sargassum by raising nutrients from deeper water layers to the surface with

45

MASLOW'S HIERARCHY OF INNOVATION –
WHAT KIND OF INNOVATION DO WE NEED?

little effort. Large machines will then compress the resulting algae mats. A portion will be harvested and processed into bioethanol as a raw material for durable bioplastics—that is, not for shopping bags that will end up floating in the ocean, but for components that will be installed in cars or computers, for example. The chemical industry giant BASF is already on board as a partner.

But it is expected that robots will compress most of the algae into something resembling large hay bales. These dense blocks of algae will then sink slowly but steadily to the sea floor several thousand meters below. This will securely bind the CO_2 for at least a thousand years. A large part will turn into sediment and be buried forever. If the technology works, this plan has the potential to be fantastically profitable thanks to CO_2 certificates. Mar Fernández-Mendéz is convinced: "Everyone's talking about trees. Of course, we can't ignore plants. But algae farms are a simpler and cheaper method to pull carbon dioxide out of the atmosphere and store it for the really long term." In addition, a major advantage of ocean-based CO_2 sequestration is that there is no shortage of ocean surface, and algae cultivation doesn't compete with agriculture. Using algae mats, it will take about 60,000 square kilometers (23,000 square miles) to remove a gigaton of carbon dioxide from the air every year. That's an area roughly the size of Croatia. On land, that would be an extremely large area, but compared to the world's oceans it's an inkblot on a nautical chart. There are still a few straits to navigate before Seafields will be able to build a facility of this size, however.

Since 2021, the London-based company has been testing the sea barriers at a prototype farm in the Caribbean, as well as harvesting, compacting, and sinking the algae bales. On a small scale in coastal waters, everything works well. But a major challenge lies in the extreme conditions on the high seas that a large-scale facility would have to withstand in order to achieve the necessary economies of

Roland Damann
SPRIND.de

scale. Fernández-Mendéz and her colleagues don't have any financial worries at the moment—in part because their team included finance experts from the start. Not many greentech investors are immune to Mar's infectious enthusiasm.[1]

While the initial steps in Seafields' unfolding story are highly encouraging, it does raise an interesting question: CO_2 sequestration using seagrass has long been under discussion as a promising method, and numerous teams from almost every continent have tried to advance the technology. The business models are plausible, the technology seems manageable, and the benefits are obvious. But many teams have seen their hopes dashed by minimal interest and money woes. Why is it so difficult to find funding and support for such a simple and intuitive idea whose successful implementation could be a highly effective weapon in our fight against climate change? Or to reframe the question in more general terms: How do we make sure that technological innovation contributes to improving the lives of as many people as possible and, if possible, not making a single life worse?

INNOVATION AND PROGRESS

In this regard, Harvard historian Jill Lepore has made an interesting semantic observation. In an essay for the *New Yorker*, entitled "The Disruption Machine: What the Gospel of Innovation Gets Wrong," Lepore describes how the terminology surrounding technological development shifted in the middle of the twentieth century.[2] The Industrial Revolution, electrification, and the great technological breakthroughs in engineering and the life sciences from the 1870s to the 1930s were generally discussed by contemporaries under the category of "technological progress." This technological optimism was

47

MASLOW'S HIERARCHY OF INNOVATION –
WHAT KIND OF INNOVATION DO WE NEED?

WHAT CAN
TECHNOLOGICAL
INNOVATION
CONTRIBUTE
TO IMPROVING
THE LIVES
OF AS MANY
PEOPLE AS
POSSIBLE AND
NOT MAKING
MAKE A SINGLE
LIFE WORSE?

always imbued with the spirit of the Enlightenment, although often naively so, as we know in retrospect. The price of progress was often high, including in the form of greenhouse gases, but in its semantic core the word "progress" implied the goal of technological advancement. Lepore referred to this goal as "betterment." Here as well, the Second World War marked a turning point.

Supposedly progress-enhancing technology had already lost its innocence in World War I's storms of steel. Beginning in 1939, it was again guilty of war crimes or at least complicit in the horror that Nazi Germany unleashed on the world with its technologically advanced weapons. The development of the atomic bomb in the Manhattan Project and the deaths at Hiroshima and Nagasaki made the dialectic of technological progress so evident that the concept of technological progress was recast by those who promoted or funded technological development. They spoke less frequently of "progress" and more often of "innovation." This probably also occurred because the economic impact of new technology as a productive factor was increasingly coming into focus not only in the United States but also in Europe and the Soviet Union. Concurrently with the Manhattan Project, Joseph Schumpeter was developing his corresponding theory of innovation at Harvard. Today, however, according to Jill Lepore's critique, the pendulum has swung too far in the other direction: "Replacing 'progress' with 'innovation' skirts the question of whether a novelty is an improvement: the world may not be getting better and better but our devices are getting newer and newer." Chapter and verse of the gospel of innovation.

In her essay, Lepore doesn't define more specifically what she means by "improvement." She only provides a comprehensive explanation of why technological disruption occurs much less often than is claimed. There is probably a simple reason for evading the question. A definition of what it means to improve the world through technology would either sound trite and at times naive, or it would be horribly difficult and complicated. Any attempt to define technological progress necessarily involves skating on thin ice. What does "better" mean, and for whom and in what sociotechnical context, taking into account any negative consequences and undesirable side effects? And which consequences should be identified as negative, and how can they be quantified? Any attempt must be based on criteria drawn from a framework of values. A framework of values can be justified in terms of the history of ideas, but since this will always be in a particular cultural context, it cannot be objective. As described in chapter 1, it is nearly impossible to assess the overall effect of technological developments in advance, especially those involving innovative leaps. The only thing that is certain is the uncertainty with which technological development is always feeling its way forward, nasty surprises included. But at the same time, it is not very helpful to demand that technology be more strongly oriented around values in the sense of a "progressive" improvement of the world without providing any value orientation.

Somewhere between naive "using-tech-for-good" rhetoric and an overly cerebral definition of progress, we propose a pragmatic approach.

The Enlightenment, and in particular utilitarian ethics, continues to provide the foundation of values for innovation conducive to advancement with and through technology. The founder of utilitarian ethics, the English jurist, philosopher, and social reformer Jeremy Bentham, summed up this benefit-oriented

THE ENLIGHTENMENT CONTINUES TO PROVIDE THE FOUNDATION OF VALUES FOR INNOVATION.

49

MASLOW'S HIERARCHY OF INNOVATION –
WHAT KIND OF INNOVATION DO WE NEED?

ethics with the "greatest happiness principle."[3] Following Bentham's principle, innovative technology must pursue the goal of creating "the greatest happiness of the greatest number." Innovations such as large-scale, deep-sea algae farms check off this box. If you measure a lot of technological developments of our time against this maxim, however, you will find that many supposedly disruptive innovations associated with the gig economy can't be linked with any kind of technological progress by definition, because the supposedly advanced business models and their digital applications maximize benefits for a few people, often at the expense of many.

Digital aggregators, such as the ride-hailing service Uber, the hotel booking platform booking.com, or restaurant delivery service platforms such as DoorDash claim to make services more usable for customers, thus increasing market efficiency. That may even be true at first glance in many cases, and who doesn't like to take advantage of the generous introductory offers for trying out these convenient services? Wherever these platform models have established themselves—often as quasi-monopolies, or at least with very large market shares—we soon discover to our surprise that the digital middlemen, ensconced between providers and customers, waste no time before exploiting their data-driven information power. The gig economy that platform models promote usually worsens working conditions. Once competition from traditional taxis has been dealt a knockout blow and demand at the airport runs high on a cold and rainy day, the platform innovation quickly turns into a very expensive proposition for customers.

Then the innovators give the high price an innovative-sounding name: "surge pricing." That's not what Bentham meant by the greatest happiness of the greatest number. If there's one thing that innovation based on utilitarian values *can't* be, it's a negative-sum game that shifts benefits or added value to a few people at the expense of

NO
RISK,

NO

LEAP

the many with the help of technology. But that is precisely the goal of the digital platform economy's pursuit of monopoly power.

Investor Peter Thiel sums it up with striking clarity in his book *From Zero to One*.[4] "Competition is for losers," according to his mantra: if company founders want to skim off value over the long term, they should try to create monopolies. The role models in his view are the superstar companies that at best create zero-sum games with their platform models and quasi-monopolistic market positions. You can understand Thiel's perspective as an investor, but you don't have to like him for it. Monopolies are very profitable for those who own them. Everyone else pays for them.

In the venture capital scene, the idea of a "foie-gras start-up"[5] has been making the rounds for some time. To make foie gras, ducks and geese are forcefully fattened for so long that their livers swell to ten times their normal size. Foie gras start-ups are stuffed full of money in order to eliminate their competition from the outset through "hypergrowth" and "blitzscaling." If the bet pays off, the result is a dominant company in a lucrative market. From the perspective of founders and investors and with good timing, this strategy can also succeed even if a dominant company fails to emerge. Because with the foie gras strategy, the idea is always to exit the venture at the right time and then leave the costs of the cleanup work to others. That's what happened at WeWork, a money-stuffed, hypergrowth start-up for shared office space.[6] Luckily, cyclic financing crunches tend to put an end to such "irrational exuberances," as then Fed chairman Alan Greenspan put it at the height of the dot-com bubble with the surefire stock market crash coming shortly thereafter.

In contrast, an agenda for innovation based on utilitarian ethics is always on the lookout for positive-sum games. Innovation is when most people win and (almost) no one loses. As is well known,

53

MASLOW'S HIERARCHY OF INNOVATION –
WHAT KIND OF INNOVATION DO WE NEED?

there is no lack of challenges where this formula can be successfully applied. These challenges have even already been systematically described.

GOALS, NEEDS, HIERARCHIES

On September 25, 2015, humanity, as represented by the United Nations General Assembly, agreed on 17 Sustainable Development Goals.[7] The envisioned target for the year 2030 rests on the three classic pillars of sustainable development: social justice, economic development, and ecological safeguarding of the foundations of life on planet Earth. The 17 goals emerge from the most pressing problems of our time and a humanistic worldview that believes in the possibility of socioecological transformation in connection with economic well-being for all.

As fully formulated, the 17 UN goals are the result of long discussions between scientists, politicians, diplomats, and those playing active roles in civil society from all continents. At first glance, none of the goals are surprising. Many of the challenges the goals address are overlapping and therefore call for cross-sector solutions developed with a focus on the complete context. There may be differing views on priorities, but one of the major strengths of the 17 Sustainable Development Goals is that they represent a robust consensus, not just an agreement on the lowest common denominator. This makes the UN Sustainable Development Goals useful, normative, and at the same time scientifically valid guides in the search for social, economic, and ecological improvement of the status quo through technology. This too is not surprising, and for that reason the goals are also frequently integrated into mission-oriented innovation agendas. The pioneers and role models in this respect are public, socially,

UNITED NATIONS
GOALS FOR SUSTAINABLE DEVELOPMENT

01 NO POVERTY

02 NO HUNGER

03 HEALTH AND WELL-BEING

04 QUALITY EDUCATION

05 GENDER EQUALITY

06 CLEAN WATER AND SANITATION

07 AFFORDABLE AND CLEAN ENERGY

08 DECENT WORK AND ECONOMIC GROWTH

09 INDUSTRY, INNOVATION
   & INFRASTRUCTURE

10 REDUCED INEQUALITY

11 SUSTAINABLE CITIES
   AND COMMUNITIES

12 RESPONSIBLE CONSUMPTION
   AND PRODUCTION

13 CLIMATE ACTION

14 LIFE BELOW WATER

15 LIFE ON LAND

16 PEACE, JUSTICE
   AND STRONG INSTITUTIONS

17 PARTNERSHIP FOR THE GOALS

55

MASLOW'S HIERARCHY OF INNOVATION –
WHAT KIND OF INNOVATION DO WE NEED?

and ecologically minded foundations focusing on innovation, such as Great Britain's Nesta, Switzerland's Innosuisse, and Sweden's Vinnova. Often the great strength of their mission-oriented approach is how it is coupled to public support through open participation in determining their mission objectives, open innovation competitions in the search for concrete solutions, and agility in adapting their goals. With NESTA and similar agencies, the structure and culture of their missions fit seamlessly into the logic of the positive-sum games described earlier. We need more of them around the world, and some politicians are having similar thoughts. To us, however, it seems helpful to have an additional classification framework for all forms of promoting innovation related to the UN's frequently cited Sustainable Development Goals.

The 17 UN goals take a bird's-eye view of collective and global problems. To supplement that distant view, it would make sense to introduce the perspective of individual benefit. For this, the five levels of Maslow's hierarchy of needs offer a classification system that is both logically rigorous and intuitively plausible. From 1943 until his death in 1970, the US psychologist Abraham Maslow formulated and continued to develop his hierarchy as a model to explain the motivations for human action in a social context.[8] Maslow was driven by the ideal of humanistic psychology with the goal of mental health through self-realization under conditions of material security. Perhaps the deeper reason why this ideal can easily be translated into a Maslowian hierarchy of *innovation* in the sense of Bentham's "greatest happiness principle" lies in the motivation of the inventor of the hierarchy of needs.

Innovation benefits individuals and communities if it achieves the following five goals:

1. Ensures basic physiological needs: clean air, water, sufficient and suitably diverse food, sleep, and the necessary conditions for reproduction.

2. Satisfies security needs: physical and mental security, basic material security, work, housing, family, and health.

3. Enhances social needs: social relationships including family, friendship, group membership and community, affection, love, and sexual intimacy.

$\rightarrow$

4. Strengthens individual needs: trust, success, appreciation, self-affirmation, independence, and freedom.

5. Enables self-realization, as the needs on the lower four levels have largely been met.

In response to our initial question as to how technology can work for human progress, we can deduce from a Maslowian innovation pyramid that technological progress necessarily helps to secure our basic and security needs. It must strengthen our social relationships and enable more freedom and independence for us, because then it will create the basis for more people with more secure needs to be able to

MASLOW'S HIERARCHY OF INNOVATION –
WHAT KIND OF INNOVATION DO WE NEED?

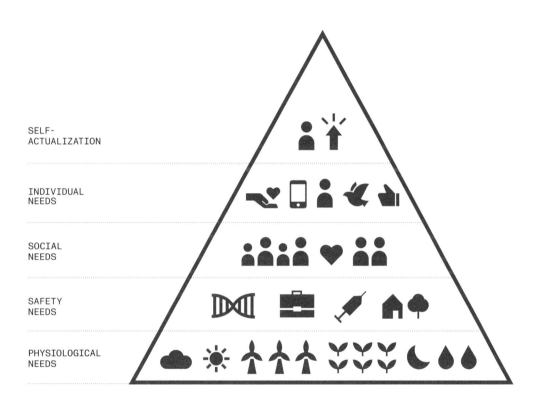

SELF-
ACTUALIZATION

INDIVIDUAL
NEEDS

SOCIAL
NEEDS

SAFETY
NEEDS

PHYSIOLOGICAL
NEEDS

strive to develop their own talents, potential, and creativity, to shape their own lives and to give them a subjectively experienced meaning.

Perhaps that sounds a little idealistic. Maybe even pathetic. But the litmus test gets interesting when applied to many recent technologies that have come onto the market under the label "social" from Silicon Valley, China, and start-up clusters in European metropolises. At what level have they secured or strengthened any particular needs, and where have they weakened the security of our needs?

Ultimately, this means that if we take a utilitarian ethic based on the humanistic foundation and values of the Enlightenment, update it with reference to the 17 Sustainable Development Goals developed in global discussions, and then align it with deeply rooted human needs, we will have a significantly better understanding of what technological progress can mean in the twenty-first century. This understanding can then function as a template. With this template, we can examine technological developments with high potential for innovation and identify candidates for technical applications where we see a high likelihood of progress for all.

The Fraunhofer Institute for Systems and Innovation Research has compiled a list of one hundred technological candidates for breakthrough innovations that seems promising to us in two ways.[9] First, the list is the result of collaboration between human beings and machines. Before the scientists took a closer look at technologies and the patents and publications associated with them, a self-learning language analysis algorithm was allowed to propose candidates. The AI for its part identified the proposals from 500,000 messages on scientific and technological platforms, with particular attention to candidates appearing for the first time.

THE MAJOR STRENGTHS OF THE 17 SDGS IS THAT THEY REPRESENT A ROBUST CONSENSUS, NOT JUST AN AGREEMENT ON THE LOWEST COMMON DENOMINATOR.

59

MASLOW'S HIERARCHY OF INNOVATION –
WHAT KIND OF INNOVATION DO WE NEED?

Second, the systematic overview provides a good sense of how broad the field is and how great the possibilities are to look for improvement corresponding to the Maslowian hierarchy of innovation. If we go through the list looking through the lens of Maslow's needs, the full spectrum of possible "betterments" in Jill Lepore's sense becomes apparent in almost every one of the technical and social innovations mentioned; may it be 3D printing of food or basic income, computational creativity or genomic vaccines, lab-on-a-chip or tidal power technologies, quantum cryptography or targeting cell death pathways, just to name a few of our favorites. We highly encourage everyone to do this exercise of the mind by scanning through the Fraunhofer Institute's list. The exercise will leave you in an optimistic mood.

Of course, real improvement requires not only an exercise of mind but that societies take advantage of these opportunities. Also of importance is how we use them.

CONTROLLED OFFENSE AND DIRTY CYCLES

Will we mount a new, miniaturized, extremely powerful laser on automated drones as a new way to kill people? Or will we use it to create a new surgical tool for Médecins Sans Frontiéres? Will AI promote the continued increase in online sales of goods produced through pollution, fraud, or exploitation? Or will the same algorithms match supply and demand on the electricity market, ceaselessly and intelligently searching with foresight in smart networks? Or, to push the question a little further, How do we ensure that innovation doesn't reinforce path dependencies by improving technology that, based on the template developed above, is bad for the general public and good for only a few people? The answers are essentially a question of society's choices and an attitude that we'll refer to as a controlled offense.

1. 2D Materials
2. 3D Printing of Food
3. 3D Printing of Glass
4. 3D Printing of Large Objects
5. 4D Printing
6. Access Economy
7. Airborne Wind Turbine
8. Alternative Currency
9. Aluminum-based Energy
10. Antibiotic Susceptibility Testing
11. Artifical Intelligence
12. Artificial Photosynthesis
13. Artificial Synapse/Brain
14. Asteroid Mining
15. Augmented Reality
16. Automated Indoor Farming
17. Basic Income
18. Biodegradable Sensors
19. Bioelectronics
20. Bioinformatics
21. Bioluminescence
22. Bionics in Medicine
23. Bioplastic
24. Bioprinting
25. Blockchain
26. Body 2.0 and the Quantified Self
27. Brain Function Mapping
28. Brain Machine Interface
29. Car-free City
30. Carbon Capture and Sequestration
31. Carbon Nanotubes
32. Chatbots
33. Collaborative Innovation Spaces
34. Computational Creativity
35. Computing Memory
36. Control of Gene Expression
37. Desalination
38. Driverless
39. Drug delivery
40. Emotion Recognition
41. Energy Harvesting
42. Epigenetic Change Technologies
43. Exoskeleton
44. Flexible Electronics
45. Flying Car
46. Gamification
47. Gene Editing
48. Gene Therapy
49. Genomic Vaccines
50. Geoengineering:
    Changing Landscapes
51. Graphene Transistors
52. Harvesting Methane Hydrate
53. High-precision Clock
54. Holograms
55. Humanoids
56. Hydrogels
57. Hydrogen Fuel
58. Hyperloop
59. Hyperspectral Imaging
60. Lab-On-A-Chip
61. Life Caching
62. Local Food Circles
63. Marine and Tidal Power
    Technologies
64. Metamaterials
65. Microbial Fuel Cells
66. Microbiome
67. Molecular Recognition
68. Molten Salt Reactors
69. Nano-LEDs
70. Nanowires
71. Neuromorphic Chip
72. Neuroscience of Creativity
    and Imagination
73. New Journalist Networks
74. Optoelectronics
75. Owning and Sharing Health Data
76. Plant Communication
77. Plastic Eating
78. Precision Farming
79. Quantum Computers
80. Quantum Cryptography
81. Read/Write Culture
82. Regenerative Medicine
83. Reinventing Education
84. Reprogrammed Human Cells
85. Self-healing Materials
86. Smart Tattoos
87. Smart windows
88. Soft Robot
89. Speech Recognition
90. Spintronics
91. Splitting Carbon Dioxide
92. Swarm Robotics
93. Targeting Cell Death Pathways
94. Technologies for Disaster
    Preparedness
95. Thermoelectric Paint
96. Touchless Gesture Recognition
97. Underwater Living
98. Warfare Drones
99. Wastewater Nutrient Recovery
100. Water Splitting

61

MASLOW'S HIERARCHY OF INNOVATION –
WHAT KIND OF INNOVATION DO WE NEED?

Artificial Intelligence and Robots

11 15 16 25 32 34 38 43 45 54 55 59 72 78 88 89 92 96 98

Society

6 8 17 26 29 33 46 61 62 73 75 81 83

Biomedicine

10 24 36 39 42 47 48 49 66 82 84 93

Electronics & Computing

30 35 44 51 53 68 69 70 74 79 80 90

Energy

7 9 12 21 41 52 57 63 65 87 95 100

Breaking Ressource Boundaries

14 23 31 37 50 58 77 91 94 97 99

Printing & Materials

1 2 3 4 5 56 64 85

Human-Machine Interaction & Biomimetics

13 22 27 28 40 71 86

Biohybrids

18 19 20 60 67 76

ISI:
100 Radical
Innovation
Breakthroughs

For technological progress, a controlled offense means that we can't react to the mistakes and misdevelopments caused by old technologies (e.g., the internal combustion engine, coal-burning power plants, petrochemicals, concrete construction, industrial agriculture, or factory farming) with "sufficiency strategies," which continue to be popular in the sustainability scene, especially in the United States and Europe. Telling people to make do with less will not lead to progress on the path toward the UN's 17 Sustainable Development Goals. The "make do with less" discourse from a rich country's perspective often comes across as arrogant. If people who live in overabundance and overconsumption call for sufficiency, it is rightly perceived as a provocation by those who haven't yet been so fortunate, notably in the Global South.

We can mount a controlled offense by taking a close look at the benefits and risks of old and new technologies with much less prejudice than in the recent past. In this, we must ask an important question: At what points can efficient technology help us find a socio-ecological transformation—including a perspective for prosperity for "the greatest number"—for a still growing world population? These also include improvements in efficiency, such as cars with one-liter engines, which, unfortunately, no major car manufacturers are seriously pursuing any longer because politics and society suddenly turned unilaterally in the direction of electric mobility. And at what points is it necessary to leap off the path that an old technology created? In the case of automobiles, that might be a fully recyclable electric or hydrogen car, or CO_2-neutral, synthetic fuels for extremely efficient combustion engines. For this, of course, not only the vehicle but also the electricity or hydrogen would have to be generated in a way that is 100 percent sustainable. Which, in turn, would require radical innovations in the field of material cycles for car production.

63

MASLOW'S HIERARCHY OF INNOVATION –
WHAT KIND OF INNOVATION DO WE NEED?

THE BEST METHOD WE KNOW TO OVERCOME PATH DEPENDENCE IS AN INNOVATIVE BREAKTHROUGH.

A controlled offense is not an either-or strategy. In fields such as motorized private transport, it pursues both lines of development—because, among other reasons, it accepts that many people in the Global South want to switch from their first bicycle to their first motorcycle and then to their first car as quickly as possible. If possible, the car should be able to drive well over 200 miles on a gallon of gas or be a very affordable electric car—as long as the electricity doesn't come from a coal-burning power plant.

In cases like the continued development of the automobile into an ecologically acceptable means of transportation, it's usually easy to reach agreement on a controlled innovation offensive. But in our experience, that's quickly no longer true when it comes to an offensive across technology minefields, such as nuclear power, genetic engineering, the extraction of raw materials, the intensive use of data, or human–machine interfaces. In these cases, the desire for control dominates in politics, science, and civil society. The possibilities for improvement through going on offense in technological development, on the other hand, are systematically underestimated. All too often, discussions on Twitter or seminar panels seem to run along the lines of, "I'd like to have my cake and eat it too." Or to put it less colloquially, there is sometimes a very distorted perception of one's own risk preferences, and decision-making premises are often prejudiced and sometimes highly ideological. The phasing-out of nuclear power in Spain, Italy, Belgium, Germany, and several other European countries provides a particularly striking example.

To be sure, climate change is slowly leading even formerly hardened opponents of nuclear power to question whether it might not have been smarter to first phase out coal-burning power generation and only then to shut down nuclear power plants step by step. But the resistance to any technological innovation related to nuclear fission or fusion is still surprisingly deep and emotional in a number of European countries. This rejection follows a pattern: Technology always causes harm. How seriously we weigh this harm is highly subjective. Each and every death in the history of nuclear power was one too many, of course. But their number is a small fraction of the deaths that coal-fired power generation causes annually in the mining industry or will have on its conscience in the long term due to its contribution to global warming.[10] In the case of oil and gas, we also have to factor in the casualties of the wars that were waged over them.

New technology can solve the problems of old technology with undesirable side effects, but only if we accept that dirty cycles can't be ruled out when it comes to radical innovations. In the case of nuclear power, the worst option is to halt innovation in the middle of the dirty cycle and argue endlessly about permanent disposal sites, even though there is clearly a good chance of making nuclear power safe and clean through further innovation and finding a solution for nuclear waste through reprocessing. History based on hypotheticals is always a game of 20/20 hindsight, of course. And yet we still presume to ask the following: Instead of a rushed phasing out (again) of nuclear energy and becoming more and more dependent on Russian gas, wouldn't it have been a lot smarter for Germany in 2011 to set up a billion-dollar fund to finance research and development related to safe and clean nuclear energy, fusion technology, and solving the nuclear waste problem? The remaining operational life spans of the current nuclear power plants could have been maintained with strict safety requirements.

Conversely, it is encouraging that more and more countries, including China, the UK, the Netherlands, Germany, and some US states, are basing their innovation strategy on a controlled offensive

65

MASLOW'S HIERARCHY OF INNOVATION –
WHAT KIND OF INNOVATION DO WE NEED?

for self-driving vehicles and allow cars with what is known as Level 4 autonomy to operate in specified areas of public transportation. Level 4 means that a human being still has to have control of the vehicle, but only has to grab the (still present) steering wheel when the autopilot no longer knows what to do and asks the human to do so. The system tries to radically reduce dangerous situations through defensive driving and emergency safety maneuvers. Autonomous driving is a prime example of how radically better technology can be introduced with controlled risk and systematic experimentation. According to the World Health Organization, around 1.35 million people die annually in traffic worldwide.[11] The world accepts this as the price of individual mobility, but if we consider the risk of driving soberly, it is obvious that there is a moral imperative to get autonomous vehicles on the road as quickly as possible for as many people as possible in order to reduce tremendous suffering as quickly as possible. One problem is that self-driving vehicles can't be exclusively trained in the laboratory, and serious accidents will occur numerous times during test drives, as has happened in isolated cases in recent years with Tesla and Uber vehicles. Tesla in particular has been repeatedly criticized for a too aggressive pace of development and relying exclusively on camera technology in order to save on the cost of expensive, laser-based lidar technology. This criticism can be made in a convincing fashion. But perhaps a shift in perspective would be helpful. How many human lives will it cost if we slow the technological development of self-driving cars by being excessively careful?

NO RISK, NO LEAP

In 1967, two years before the first moon landing, the US Senate's Science and Space Committee called a hearing to discuss possible harm to the human body from supersonic flight. Experts spoke, including skeptics who warned that the lack of studies meant the risk was incalculable. The speakers also included physicians, who saw little risk of consequences to the human body from supersonic flight. As it turned out later, the dangers of extremely fast aircraft have more to do with crash landings than with acceleration during takeoff or the shock of breaking the sound barrier. In the history of science, however, the committee meeting is important for an entirely different reason. For the first time, the invitation contained a term that was just then gaining attention in the US scientific community but had not yet reached the political arena: "technology assessment." [12]

At its core, technology assessment revolves around a dilemma that David Collingridge, a British scholar of technology policy, was first to express precisely and in its full significance in his 1980 book *The Social Control of Technology*. [13] The economic, social, and ecological consequences of innovations, both positive and negative, can only be measured and assessed after a technology has spread. We've already made note of that. But to make matters worse, "technology assessment" often only provides the desired information when technology has already advanced so far that the wheels of technological history can no longer be rolled back. It is either too expensive, or society can no longer reach a consensus because too many technology users value its benefits too much. For understanding the Collingridge dilemma, the automobile offers an instructive example. On the one hand, cars are far too dangerous and environmentally harmful to stand a chance of obtaining regulatory approval today. On the other hand, no society is likely to accept the costs of abolishing them, either economically or in terms of individual mobility.

67

MASLOW'S HIERARCHY OF INNOVATION –
WHAT KIND OF INNOVATION DO WE NEED?

The essence of the dilemma is that there are no conventional solutions to it. Innovation research uses the term "exnovation" to describe the phasing out of technologies that in retrospect turn out to be harmful.[14] Exnovation also has high follow-up costs. This is why exnovation is so difficult for us, and not just in the case of the automobile. The best method we know is an innovative breakthrough that overcomes its own path dependency.

A central question of our time for politics and society is which risks we're prepared to take in order to find potentially radically better solutions. It's not a binary decision, of course. It always involves a difficult balancing of factors that has to be made on the basis of incomplete information and under conditions of heightened uncertainty. If we knew that autonomous vehicles would cause only 500 deaths each year, but the thousands of victims of human-driven cars would survive, the decision would be easy. But we usually don't have this information in the form of statistical certainty.

The strategy of the controlled offense presumes that innovators will experiment as much as possible and gauge the potential consequences of technology. However, we often need to remember the price of caution as well. In pharmaceutical research, it's becoming especially clear how fine the line is between the necessary assessment of consequences and the cost of excessive caution. Since the 1960s, research-based drug manufacturers have been required to demonstrate in great detail that any new medications have minimal side effects before they can be introduced.

Historically, the radical tightening of regulations for assessing drug risks was the correct and necessary reaction to the horrific consequences of the approval of thalidomide. It was initially even available without a prescription. Thalidomide led to severe birth defects in many thousands of newborns worldwide. Recent estimates assume more than 24,000 victims born alive. The number of miscarriages and stillbirths might be well over 120,000.[15] The burdens of the families affected are, of course, unbearable. Financial compensation has often been, to put it mildly, modest. But the costs resulting from the radical tightening of the approval process according to the principle of nonmaleficence are also an open secret in the pharmaceutical industry. If a newly discovered substance works particularly well in early laboratory tests, it's very often rejected from further consideration. As a rule, high potency is usually paired with especially severe side effects. A pharmaceutical manufacturer then faces a high risk of being unable to bring the side effects sufficiently under control, despite years of development and billions in R&D costs, to eliminate any safety concerns. We don't know how much innovation we've missed out on over the past fifty years because of this excessive caution.

What does all this mean for us? It's risky not to take risks. We can increase our awareness of when it's worth going on offense to develop technology to help us escape from paths leading in the wrong direction by making giant leaps. Our goal is the greatest happiness of the many, and the UN Sustainable Development Goals, coupled with a close look at human needs, point in the direction we should leap. Public debate must constantly reflect what the right direction for innovation leaps could be in any given context. And independent oversight, based on scientific data, must be a built-in part of any responsible innovation policy and strategy. This is very obviously not an easy feat but well worth the effort for all societies that want to push innovation for progress. Despite the uncertainty about the benefits and harms of new technology, at least one thing is clear: those who in their excessive caution fail to drive innovation forward only leave innovation in the hands of others and consequently have less influence on the form it takes. So we need to identify and support those people who are both courageous and spur development responsibly. There are many of them.

69

MASLOW'S HIERARCHY OF INNOVATION –
WHAT KIND OF INNOVATION DO WE NEED?

The next chapter tells how the people who make innovative leaps see the world, how they think, what makes them tick, and how they live—and what they need to drive technological progress forward.

THE POSSESSED

WHO MAKES INNOVATION LEAP?

THE FUNGUS NERD

The year 1948 was a good one for the life sciences. The US Congress and President Harry Truman cleared the path and approved the funding for the National Heart Institute. The same year saw the launch of what is still the largest and probably most impactful long-term population study in medical history. Not far from Boston, in the small town of Framingham in Massachusetts, more than 5,000 citizens agreed to give doctors extensive information about their life circumstances and habits—and to receive regular medical checkups.[1] The aim of the Framingham Heart Study was and is to identify risk factors for heart attacks and strokes. Before the study, medical science had determined that heart disease was the leading cause of premature death in the United States. The number of cases was steadily increasing. But the life scientists were left guessing whether obesity, smoking, and alcohol consumption increased the risk of heart disease, and if so, by how much—and whether and to what extent physical exercise or other aspects of a healthy lifestyle reduced the risks. For one central question, science had only vague hypotheses: What kind of food leads to heart disease? In 1961, after thirteen years of observation, the lead scientists published a series of findings that soon became the medical standard. High blood pressure and elevated blood sugar had long been in discussion as risk factors. This was now empirically confirmed. But to the surprise of many researchers, the Framingham study identified high cholesterol as one of the most acute threats to heart health.

It took these new discoveries around ten years to have a statistically measurable public health effect on Americans' behavior and lifestyle. Through the end of the 1960s, the rate of coronary heart disease continued to rise. Since then, it has fallen continuously and is now per capita around a third of its peak rate half a century ago. A healthier diet, more exercise, and less smoking have all contributed to this encouraging development. But the contribution of statins, drugs to reduce cholesterol, is likely even more significant. Sales of statins are among the highest of all drugs today. The American geneticists Michael Brown and Joseph Goldstein received the 1985 Nobel Prize in Medicine for their work on cholesterol metabolism and statins. It was certainly deserved. But when the prize was awarded, the work of a third scientist was unfortunately ignored. The basis for the statin breakthrough had actually been created in the early 1970s by a Japanese biochemist, the child of a peasant family who, already as a young boy, was highly interested in fungi and the history of science.[2]

In 1966, five years after the first extensive publications based on the Framingham study, thirty-three-year-old Akira Endo came to Albert Einstein College of Medicine in New York and spent two years there as a visiting scientist. At the time, Endo was working for the pharmaceutical division of the Japanese conglomerate Sankyo. Einstein College was one of the hotspots of cholesterol research. People had high hopes that this new line of research could find a solution to heart attack, a disease that killed countless Americans every year. The research accordingly enjoyed generous funding in the United States, and the first large studies with cholesterol-lowering drugs were underway. Interest was also rising in Japan, because Western eating habits gradually spreading there were leading to an increase in heart disease. In New York, Endo was quite surprised at how many people looked like sumo wrestlers. As for the two years of paid research in New York, they were a reward for the biochemist's outstanding achievements. Working in his employer's lab, Endo

MUSH-ROOMS CAN'T RUN AWAY FROM THEIR ENEMIES. SO EVOLUTION HAD TURNED THEM INTO OUTSTANDING CHEMISTS.

had previously identified a mold that was useful for the food industry, *Coniella diplodiella*. It produces certain enzymes that extend the shelf life of fruit juices. Sankyo had achieved tremendous commercial success with the process thanks to Endo's interest in fungi. It was the story of Alexander Fleming's accidental discovery of the penicillin mold's antibacterial effect that originally caught his interest. Even as a schoolboy, Endo had attracted attention with an experiment in which he extracted a substance from mushrooms that killed flies but was harmless to humans. He grasped early on that mushrooms can't run away from their enemies. So evolution had turned them into outstanding chemists. In New York, the Japanese scientist learned that many bacteria require cholesterol for survival. From this, he hypothesized that perhaps some fungi had evolved to form a chemical reaction against cholesterol. The person who found such a mushroom would have a natural weapon for fighting cholesterol. Back in Japan, he went looking for it.

In Sankyo's laboratories, Akira Endo first spent three years developing an intricate testing apparatus. Beginning in April 1971, he let 6,000 types of fungi do battle with cholesterol-producing bacteria in petri dishes. Over a year later, he found something: a blue-green mold found in a Kyoto rice shop blocked one of the key enzymes in bacterial cholesterol production. The fungus, *Penicillium citrinum*, was a cousin of Fleming's penicillin mold. After another year of work, Endo had extracted the crucial anticholesterol molecule from the fungus and named it ML236B. This molecule eventually became the progenitor of all the statins that today save millions of lives every year. But Endo's experiment almost ended in utter failure.

The timing of Endo's discovery couldn't have been worse. The usual testing and approval procedures were an awkward fit for ML236B. And Endo's career and reputation were now at stake to such an extent that many other scientists would have given up. In retrospect, the history of statins provides a textbook example of the characteristics and skills a person has to have to make an innovative breakthrough.

When the Japanese scientist came to New York in 1966, the research community was extremely optimistic that a cholesterol-lowering drug would soon be found. As he was making good progress with his fungus experiments only half a decade later, the tides of science had already turned. Several studies had in the meantime reached the conclusion that low-cholesterol diets had no effect on heart disease. A bold claim soon became the consensus opinion among biologists, pharmacologists, and biochemists: any cholesterol-lowering agent would necessarily be dangerous because cholesterol is essential for core cellular functions. Depriving cells of cholesterol would therefore completely disrupt the cells' biochemical processes.

The search for cholesterol-lowering drugs had gone from being a beacon of hope to a hopeless case for researchers who hadn't gotten the scientific memo and obviously didn't understand the foundations of cellular functions. When Endo presented his results with *Penicillium citrinum* at conferences in the mid-1970s, the lecture hall would quickly empty. He was also meeting increased resistance from his employer, Sankyo. For years, the company had given the innovator broad leeway to conduct research on mold, but now the one-time discoverer of a juice-sterilizing enzyme seemed to have gotten off on the wrong track. His supervisors finally wanted to see usable results, while his career-focused colleagues were increasingly distancing themselves from him.

Akira Endo mentally steeled himself to be fired. He made an agreement with his wife that, if necessary, she would support the family on her own for a while. He wrote a letter of resignation that he carried with him constantly from that point forward. If he was asked to abandon his research, he would finish the matter with his head held high. But to his own surprise, Endo was allowed to continue his work a little longer, thanks to one last supporter in Sankyo's innovation management, and test the substance he had extracted from mold on a few rats in the company's testing lab. The result couldn't have been clearer: it had no cholesterol-lowering effect. In retrospect, that was the lowest point. The researcher should actually have given up, because such a clear result on a standard test involving rats mandated the end of the Sankyo development process. That is, unless two researchers *break* the rules of the standard development process.

Over an after-work beer in a bar near the labs, Endo met a colleague from another department who was doing experiments on chickens. They wondered whether chickens might be particularly good test animals for cholesterol research because their eggs contained so much cholesterol. Without permission, the colleague gave a few chickens to Endo, and a few weeks later they both knew that ML236B greatly reduces cholesterol levels in chickens. The birds showed no sign of side effects. The results were so unmistakable that they even convinced the skeptics in Sankyo innovation management. Endo was allowed to continue testing on dogs and monkeys and finally published a scientific paper in 1976 that caught the attention of two physicians at the University of Texas: Michael Brown and Joseph L. Goldstein, the later Nobel Prize winners. The duo requested a sample of the

ONLY A FEW SUBSTANCES IN THE HISTORY OF MEDICINE HAVE SAVED AS MANY HUMAN LIVES AS STATINS.

substance from Endo, which the Japanese scientist immediately provided.

In laboratory tests in Texas, the original statin had the same desired effect as in Japan, and again without side effects. Brown and Goldstein immediately wrote a recommendation for Endo, which paved the way for him and a medical researcher to conduct a first test in Japan in 1978 with a patient who was at severe risk of heart attack. Shortly afterward, Sankyo began a study involving twelve Japanese hospitals. Their results caused a worldwide sensation in 1980. However, the approval process for statins again proved to be a rocky path—among other things, because of an improperly conducted study that had found a risk of cancer in dogs. However, Akira Endo's biochemical innovative leap was finally on the right track. Leading physicians around the world were convinced and took part in the clinical development. Major pharmaceutical companies had jumped back on board with cholesterol-lowering drugs and gotten the regulatory authorities involved. Endo himself was no longer able to accelerate development. He accepted a professorship in agriculture and technology at Tokyo University. In 1987, almost four decades after the start of the Framingham study, US regulatory authorities approved the first statin, Mevacor, from the pharmaceutical company Merck. Only a few substances in the history of medicine have saved as many human lives as statins. Brown and Goldstein, of course, had no influence on the Nobel Prize committee's decision to ignore the Japanese innovator. In 2004, they dedicated a text to him with the title "A Tribute to Akira Endo, Discoverer of a 'Penicillin' for Cholesterol." Their conclusion: "The millions of people whose lives will be extended through statin therapy owe it all to Akira Endo and his search through fungal extracts at the Sankyo Co."

THE HIPOS PSYCHOGRAM

Outside Japan, Akira Endo's story is less well known than that of many other great innovators. That's a shame in several ways. An innovator who has had so much positive impact on so many people's lives deserves to be known. Endo's success can inspire future innovators. This is one reason we tell his story here in such detail. Above all, however, Endo's biography illustrates with particular clarity the skills, character traits, and attitudes that the people who bring about innovative leaps need to have.

But first, a scientific caveat: To our knowledge, there is no data-based research on the "psychogram" of people who make innovative breakthroughs. At most, there are studies on partial aspects. This includes the role of experience gained over many years as a prerequisite for turning ideas into fundamentally better solutions. But even here, the science is unclear, often contradictory, and by no means reducible into a simple "10,000 hours of hard work rule," as the Canadian author Malcolm Gladwell proposes in his worldwide bestseller *Outliers* [3] with narrative brilliance, but on a questionable empirical basis. We too can suggest an approximate answer to the big question, How do great innovators really tick and collaborate? only with anecdotal evidence.

On the following pages, we will refer to the shelves full of books that derive striking patterns and factors for successful innovation from the life and work, motivation and working conditions, values and victories of great scientists and technology entrepreneurs. In addition to the biographies of great innovators from Marie Curie and Robert Bosch to Steve Jobs and Elon Musk,[4] the following current overviews of the field are particularly worth reading: *Originals: How Non-conformists Move the World* by Adam Grant; *How to Fly a Horse: The Secret History of Creation, Invention and*

Discovery by Kevin Ashton; *How Innovation Works: And Why It Flourishes in Freedom* by Matt Ridley; and *Loonshots: How to Nurture the Crazy Ideas That Win Wars, Cure Diseases and Transform Industries* by Safi Bahcall, who is himself a pharmaceutical innovator and gives a comprehensive treatment of Endo's story in his book.[5]

In addition, we can support our approximate psychogram of great innovators with dozens of in-depth interviews and intensive discussions conducted by Thomas in his regular SPRIND podcast and by Rafael and the innovation managers at the Federal Agency for Disruptive Innovation in their search for so-called HiPos.[6] People with a high potential for radical innovation almost always have the characteristics displayed by Akira Endo as he pursued his innovative leap from an intuitive idea to a blockbuster drug. Intelligence, a talent for abstraction, and combinatorial thinking are part of it. But these are more like hygiene factors—necessary but not sufficient prerequisites that many talented people and top performers fulfill.

Those who make innovative leaps, however, usually show high levels of five additional personality traits or behaviors:

1. Early specialization: an interest, barely comprehensible to others, in a specialized area, sometimes bordering on manic obsession

2. Grit: an unusually high level of tenacity and resilience in the face of setbacks, combined with steadfast independence of mind in response to criticism or even ostracism

3. The desire for one's own work to have a genuine impact

4. The ability to inspire and empower: to transmit their own enthusiasm to others and to build and lead teams without descending into micromanagement

5. Having an impact: the desire for their own work to actually have an impact

$\longrightarrow$

1.
EXTREME
INTEREST

2.
GRIT AND
TENACITY

3.
OPENNESS

4.
EMPOWERING
LEADERSHIP

5.
DRIVE FOR
IMPACT

1.

Early specialization

The five points overlap and reinforce each other. On the surface, some of them may sound like nerd clichés of the type found with increasing frequency in popular culture in recent years, often ironically, for example, in *The Big Bang Theory* television series. The quality of pop culture does increase, though, with the producer's ability to observe reality accurately; perhaps Endo could have provided the template for an interesting character as the congenial counterpart of Sheldon Cooper, the stereotypical physics genius.

Akira Endo grew up in a large family on a farm in northern Japan. His grandparents also lived on the farm. His grandfather transmitted his deep interest in medicine in general and mushrooms in particular to the young Endo. In a short, modest autobiography for the Japan Academy, Endo emphasizes that his interest in the chemical processes in mushrooms was awakened at the age of ten and that it became the focus of his life at the latest after he devoured a biography of Fleming. He also developed a desire to become a great scientist while he was still a child. The HiPos at the Federal Agency for Disruptive Innovation often report very similar childhood experiences. A mentor, often a close relative or a particularly inspiring teacher, introduced them to a topic, whether it was electronics, calculating machines, biochemistry, mechanics, physics, or environmental technology. They were never able to let go of the topic, and specialization often took place at a very early age at the expense of other interests or the willingness to deal broadly with other topics. With HiPos, it is almost considered a membership requirement to have never been shortlisted for elite academic scholarship for gifted students. Their failing grades in French or social studies or gym always got in the way. Many HiPos also have the feeling that scholarship holders are expected to be social conformists. Maybe other people with straight-A high school GPAs are capable of and interested in meeting that expectation, but they themselves are not. It would certainly be foolish to deny that broadly gifted, hardworking, and socially skilled young people have the potential to make innovative leaps. But it is striking that many innovators—both historically and today, from Einstein to Jobs to Musk—have exactly the kind of uneven educational record that suggests an excessive, nerdy interest in a specialized area—and the courage to accept gaps in areas that don't interest the young nerd at all. Dropping out of the university can also be an encouraging indication that a young person wants to make an impact through innovation—and possibly might even do so. This pattern is found not only in the biographies of many successful university dropouts in Silicon Valley but increasingly among European and Asian start-up founders as well.

Certainly, there's a place in education policy debates for the never-ending dispute among educators over how much specialization at the expense of a broad education as defined by the traditional canon is appropriate, and at what age. In innovation research as well, there are renewed calls at regular intervals for the revival of the "Renaissance person," or the significance of comprehensively educated generalists in the innovation process.[7] But if you ask the people behind breakthrough innovations, their self-reflections repeatedly emphasize first, that their early specialization was a prerequisite for their path to becoming an innovator, and second, that specializing to the desired degree often required them to overcome the stubborn resistance of their parents or school, especially since no one understood why a young person would be so interested in a particular area that they could pursue their interest at the cost of their general education. Many HiPos emphasize that in the long term, this struggle was productive. At least in their perception, it strongly contributed to the second personality trait of innovators: an unusually high degree of tenacity and endurance. Akira Endo is a good example of this too.

 # 2.

Grit

"It's not a good drug unless it's been killed at least three times." The sentence comes from the British Nobel Laureate in Medicine Sir James Black, who received the award in 1988 for his work on the biochemical principles of drug therapy. Endo's development of statins from mold was on the verge of failure three times. After returning from the United States, the Japanese innovator found little support from Sankyo because the majority opinion of the scientific community had turned away from research on cholesterol. Someone less tenacious than Endo would probably have chosen to apply his skills and knowledge to another research question. A researcher without independence of mind wouldn't have gone to work at the lab for years on end with a letter of resignation in his pocket so that he would be ready at any moment to tell his supervisors in the hierarchical and identity-defining Japanese corporate culture of the 1970s, "Then I'll just go develop my idea somewhere else!"—while knowing full well that it wouldn't have been easy to find that "somewhere else."

 TED Talk
Angela Lee
Duckworth:
"Grit"

    After the failure of the efficacy study with rats, most researchers would probably have given up. But ultimately, even the dogs who supposedly developed cancer from statins had to accept their effectiveness. The phrase "cancer causing" is usually the death

"IT'S NOT
A GOOD
DRUG
UNLESS
IT'S BEEN
KILLED
AT LEAST
THREE
TIMES."

knell for a project in the world of pharmaceuticals. As a professor at the University of Tokyo, Endo took the opportunity to refute the flawed study. This was the third resurrection of his idea. It took roughly twenty years from the hunch that a fungus species could be a chemical weapon against cholesterol to the approval of Merck's Mevacor. At this point, Akira Endo was in his mid-fifties. The American psychologist and neuroscientist Angela Lee Duckworth would say that Endo had shown "grit."[8]

Duckworth has repeatedly emphasized the aspect of perseverance, for example, in her book *Grit: The Power of Passion and Perseverance*. Her book summarizes a large number of studies demonstrating that perseverance is a better predictor of long-term success than intelligence, talent, or social skills, such as a knack for communication. In her view, this applies to professions as diverse as musicians, mathematicians, managers, and Navy Seals. Unfortunately, to the best of our knowledge, there are no empirically validated studies on the correlation between the success of innovators who make breakthroughs and tenacity, which would probably be difficult to carry out in any case. It's impossible to conduct systematic in-depth interviews with Marie Curie, Robert Bosch, and Steve Jobs posthumously or measure various characteristics with personality tests beyond the grave. In our conversations with HiPos, however, "grit" clearly stands out as one of their most striking personality traits.

Many promising candidates for funding from the Federal Agency for Disruptive Innovation aren't young scientists but people in their fourth or fifth decade, sometimes even significantly older. The oldest just turned ninety-two. Horst Bendix, who designed cranes and coal mining excavators for decades in East

EXCELLENCE REQUIRES MUCH MORE PERSISTENCE THAN THE MANY STORIES OF CHILD GENIUSES AND OVERACHIEVERS SUGGEST.

Germany, is now developing a giant wind turbine with a delicate, tubular steel construction. As previously mentioned, Malcolm Gladwell's "outlier" rule, according to which superior knowledge and ability arise after around 10,000 hours of intensive work in a field, has been criticized as an invalid oversimplification by the authors of the scientific study that Gladwell based his rule on. The Swedish American psychologist Anders Ericsson accused Gladwell of having chosen this threshold because it offered a nice round number;[9] Gladwell then bent the anecdotal examples of Joseph Oppenheimer, the Beatles, and Bill Gates to fit 10,000 hours. For Ericsson, however, the general point of Gladwell's claim remains beyond dispute: excellence requires much more persistence than the many stories of child geniuses and overachievers suggest. There is almost no area where excellence can be achieved on talent alone. For most HiPos working toward an innovative breakthrough related to technology, however, the 10,000-hours-rule is irrelevant anyway. Their accumulated experience in their field is usually significantly higher.

If you ask HiPos about their "grit," they often don't immediately understand why the question is being asked; for them, perseverance and tenacity have nothing to do with the self-discipline described by star athletes in sports that require intense training. The HiPos' stamina is rooted instead in their deep interest. They're more likely to have to exercise discipline if they're forced to deal with something that's outside their area. Bernd Ulmann, an innovator behind analog computers and a prototypical HiPos for the Federal Agency for Disruptive Innovation, describes it as follows: "The day has 16 hours, and every hour that I don't spend on my computers feels like an hour lost."[10]

The term *work–life balance* doesn't occur in Ulmann's vocabulary. Work *is* life. Period. Ulmann knows it sounds like a cliché, but that's just the way it is. There's certainly variety in his working life, however. As a "mental load shift," Ulmann recently built a gamma spectrometer to measure a particular kind of radioactive ray, which "really has nothing to do with analog computing." The term "load shift" sounds like how an athlete would describe training minor muscle groups in order to push the limits of the main muscle even further. Crossing boundaries is the nature of innovative leaps.

What links the computer scientist Bernd Ulmann with many of those who make breakthrough innovations is a deeply rooted conviction that they see things in their field of expertise that others have overlooked. In conversation, this type of self-confidence tends to come across not as arrogant but as neutral distancing. The outsiders are those who have never focused on a topic in comparable depth. It is precisely this attitude that makes breakthrough innovators—and Akida Endo is once again a good example—mentally impervious to criticism or hostility from others. HiPos often perceive criticism from the superficially informed as confirmation that they must be on the right track; their motto is, "It's precisely because no one but me (and a few comrades-in-arms) can see it that there's an opportunity for a breakthrough innovation here." That's logical. Otherwise someone elsc would have long since made the discovery.

3.

Openness

Having a thick skin when it comes to destructive criticism or even personal hostility shouldn't be confused with a lack of openness to constructive criticism or valuable impetuses from outside that are capable of accelerating one's own innovation. This fundamental openness in both senses—sharing your own knowledge and absorbing interesting ideas—requires careful examination because in popular narratives about the great inventors, the importance of this quality is often underestimated, ignored, or even denied. The biographies of the great innovators in the history of technology often read like the heroic stories that were written when the great man theory of history still prevailed. The Scottish historian Thomas Carlyle formulated this theory in the nineteenth century and—as the name suggests—used it to interpret the progress of world history as the result of extraordinary actions by great men who did the right thing in the right place at the right moment.[11] On top of that, they were born as great leaders (since according to the great man theory, leadership skills can't be learned). Starting with its name and its exaggerated masculinity, we think the great man theory is a nonsensical explanation for the general progression of events over time and especially for the history of technology.

The great majority of our conversations with breakthrough innovators don't leave the impression of isolated geniuses, similar to the Russian mathematician Grigori Perelman, who hoped to contribute to world progress through technology. Perelman avoided interacting with the rest of the world so strictly that he rejected both the Fields Medal, the most prestigious award in mathematics, and a million dollars in prize money from the Clay Mathematics Institute for his solution to the Poincaré conjecture. In contrast, the majority of those who make breakthrough innovations are especially open to exchanging ideas with anyone who can help them sharpen their thoughts. Yes, they are resilient whenever anyone, for whatever reason, wants to stop them. The same applies to setbacks that they don't understand at first, but for which there must be an explanation and a solution. In nerd culture, openness to constructive criticism—as well as the willingness to invest time in the constructive criticism of other projects—is a widely accepted virtue. Successful innovators embody this virtue, and we wonder why the innovator-as-hero epics regularly skip over it. Remember Akira Endo. His test rats didn't respond to statins because their veins are flowing with HDL, colloquially known as "good cholesterol," but very little LDL, or the "bad cholesterol" that causes heart disease. For that reason, the fungal extract had no effect on them. Chickens and human beings, on the other hand, have both types of cholesterol. The decisive clue that led to perhaps the most important breakthrough in the development of statins was discovered while Endo was shooting the breeze over an after-work beer with a colleague who worked with chickens. It's highly unlikely that Endo would have come up with it on his own. His colleague not only provided the key insight while sharing a beer, he also granted quick and unbureaucratic access to test chickens by flexibly interpreting corporate rules. This type of informal support with a major impact on highly impactful innovations is again not an isolated case but a method in itself.

In her classic 1994 study *Regional Advantage*, the innovation researcher AnnaLee Saxenian, professor at UC Berkeley's School of Information, perceptively and extensively described this culture of exchange and mutual support as a central factor in the rise of Silicon Valley to the most successful innovation cluster in the world.[12] It wasn't (only) the terrific weather in Northern

THE
SAVE

NERDS
THE
WORLD

California that attracted the best tech talents of the United States—and soon the whole world—to the Valley starting in the 1960s and with increasing pace in the 1970s, but a new answer to the question, How should we work (together)?

The West Coast start-up culture, with Stanford as a training center for new talent, was built on information flows not only within teams or in companies with flat hierarchies but also between companies, even if they were sometimes in direct competition with each other. The engineers and employees of various start-ups met after work in cafés and wine bars and exchanged ideas with an openness that would have been inconceivable at what had been the dominant tech hub up to that point, Route 128 in Boston. Until the early 1970s, this high-tech highway in the vicinity of MIT and Harvard had been the United States' hotspot of digital technological breakthroughs. Digital Equipment Corporation (DEC), radar and microwave specialist Raytheon, instant camera pioneer Polaroid, and the early mini-computer manufacturers Data General and Wang had all risen to prominence here after World War II. By the end of the 1960s, there were more than a thousand high-tech companies on Route 128. But then this success story took a sharp turn off course.

Saxenian describes in striking fashion how treating knowledge as an instrument for internal dominance and an exaggerated sense of secrecy toward outsiders turned into structural disadvantages for the vertically integrated and hierarchically organized tech companies on the East Coast compared to their up-and-coming regional rivals on the West Coast. In Silicon Valley, innovations made at Xerox PARC were inspiring Apple's Macintosh, and chip developers were granting each other the right to use their patents. But on Route 128, one company after another was going bankrupt or being sold off at fire-sale prices. The closed-shop mentality in the East couldn't keep up with the collaborative attitude of the West Coast innovation process. Here again, the example of how statins were discovered teaches an enduring lesson. Endo constantly shared his knowledge, regardless of his own losses. As the saying goes, knowledge is a resource that increases with use. That researchers in the United States were able to keep development moving forward was more important to him than his own career or Sankyo's potential profits. Whenever scientists with access to more or different resources than he had would ask him about the current state of his research, Endo would share data and knowledge generously, well beyond the usual level found in scientific publications. And he was capable of letting go of his innovation—and thus control over his life's work—at the moment when it became clear that he himself could no longer make a significant contribution to it. He let others keep moving forward.

Of course, not all innovators have maximum openness and cooperation as their default settings. And even if they do, they don't always meet up with exactly the right people for an inspiring exchange of ideas. An intriguing approach to publicly funded innovation involves promoting this type of exchange or even compelling a degree of cooperation under competitive conditions. The principle is simple. Several teams are financed and directed to answer a particular question or complete a certain mission. To a certain extent, whoever accepts public money has to share their knowledge with competing teams.

Perhaps the most striking example of this form of government-induced "co-opetition"—a mixture of cooperation and competition—is the development of the mRNA COVID-19 vaccine. The role of DARPA in this success story is surprisingly

WITH FEW
EXCEPTIONS,
INNOVATION
TODAY IS A
TEAM SPORT.

94

CHAPTER 3

little known. In 2013, innovation manager Dan Wattendorf launched an mRNA program in which around ten teams took part and were forced to exchange information each month.[13] In personal conversation, CureVac founder and mRNA pioneer Ingmar Hoerr (among others) has vividly described how "painful" this sometimes was. But the exchange of information was also extremely successful. The ten teams participating in 2013 included the breakthrough innovators from Moderna and BioNTech. Dan Wattendorf seems to have done a lot of things right selecting the teams and designing the collaborative-competitive development process. He often stepped into the role of moderator between the innovation team leaders. He was a coach of the coaches, so to speak, which leads directly to the next, perhaps most important, skill set of breakthrough innovators.

$\rightarrow$ ## 4.

Ability to Inspire and Empower

With few exceptions, innovation today is a team sport. This leads to a seemingly simple question that we constantly encounter in all possible contexts of innovation discourse, both theoretical and practical: To what extent do breakthrough innovators have to be team players? The question has also repeatedly come up at the Federal Agency for Disruptive Innovation (SPRIND). But more and more often, we're getting the impression that this is the wrong question. The question actually needs to be, What qualities do breakthrough innovators need to have in order to create and lead teams that are capable of finding solutions that are radically better than what came before? Here again, hard empirical evidence is difficult to find. But what's striking about many proven breakthrough innovators, such as BioNTech cofounder Özlem

Türici, 23andMe cofounder and CEO Anne Wojcicki, or the usual suspects such as Steve Jobs and Elon Musk, is that they're leaders whose enthusiasm for their cause is almost irresistibly contagious for anyone interested in technology and progress. As we see it, it's not a matter of natural charisma (as the great man theory would postulate). Özlem Türici, for example, is reserved to the point of shyness. And if he was being scored for charisma, Mark Zuckerberg would hardly have been able to revolutionize social interaction on the Internet. The same is largely true of soft-spoken innovators like Sergey Brin and Larry Page, who threw open the doors to knowledge with Google, Vinton Cerf and Bob Kahn, who created the foundations of the internet with the TCP/ IP protocol. But when Türici speaks about mRNA with quiet expertise, or Cerf talks about the basics of programming as the basis for a better world, you can't help but fall under their spell.

Our experience with the selection of HiPos for funding through SPRIND is also that the contagiousness of the enthusiasm for a new technology will match the innovator's inner conviction of their own vision and mission. Naturally, this plays a major role in securing capital, but for the mission's chances of success, it's more crucial that other highly gifted people can be infected with the innovator's enthusiasm. That's only possible if the breakthrough innovator has the necessary deep expertise. Investors may be deceived now and then by glib phonies. But it's not possible to deceive people who are themselves deeply involved in the topic—and they're the ones who are the best potential colleagues for the undertaking. If new team members detect a lack of competence on the founder's part, they won't stick around. In addition to deep conviction and expertise, however, a third skill is needed. It's prominently mentioned in almost every modern textbook on leadership, but it's often a challenge for innovators in the "nerds with a mission" category: They have to give the other nerds on the team sufficient freedom and empower them to push the innovation process forward collaboratively. This is often where the wheat is separated from the chaff—that is to say, people who have great ideas and are technically brilliant, but unfortunately aren't the kind of leaders who can empower others and thus fail to reach their potential as breakthrough innovators.

At SPRIND, it's sometimes almost tragic to see how the projects of brilliant minds fail because, as highly talented specialists, they can't get past the principle of command and control. On the one hand, this can be explained psychologically. Their extreme interest in a particular field in which they themselves are towering figures naturally makes it difficult for them to delegate responsibility. Then the HiPos' penchant for micromanagement scares off exactly the people these innovators had previously inspired with their vision and attracted to their undertaking—or precisely the people best able to move things forward as members of the team. And conversely, the chances of success increase when strong, capable tech founders increasingly take on the role of representing their project to the outside world and securing attention and capital, but giving other strong personalities in the second tier the necessary freedom to make decisions about technological development and implementation. Think Bill Gates, Steve Jobs, and (of course) Elon Musk. A good example of empowering leadership can be seen in Gwynne Shotwell, the engineering mastermind and the entrepreneurial driving force behind SpaceX. What from the outside often looks like innovation based on the great man theory is actually the result of a great leadership and team model.

→ # 5.

Having an Impact

What drives breakthrough innovators? As an answer to that question, Paul Graham developed the "Bus Ticket Theory of Genius."[14] Graham is a programmer, venture capitalist, and cofounder of Y Combinator, the legendary Silicon Valley start-up incubator. He knows a lot of nerds. The starting point of his bus ticket collector theory is the deep interest in a specialized area that goes far beyond what others can comprehend. His models for this are the kind of obsessed people who deeply and endlessly occupy themselves with different bus companies' ticket specifications, route networks, and tariff systems. Most people find this to be just as useful as the ability to recite the order of batters on both teams in game seven of the 1982 World Series between the Brewers and the Cardinals and, of course, to know who got a hit and who didn't. For Graham, the difference between bus ticket collectors and breakthrough innovators is not just that genius innovators are obsessively interested in something useful, or that at least could become useful. Breakthrough innovators are above all prepared to convert their obsession into something useful "when the opportunity arises." This is exactly where he hits on the crux of the matter.

Many scientists all over the world, have an obsessive interest in their field, which is of course a prerequisite for academic excellence. But their obsession leads relatively few of them to develop the additional ambition to have an impact beyond citation indexes. This is especially true for Europe. In terms of number of spin-offs from science-related companies, Europe lags far behind North America and tech-ambitious Asian nations.

Don't get us wrong: of course basic research without immediately obvious practical application has a lot of value. It's important and correct, and scientific freedom includes the freedom to research without pursuing any other aim than satisfying curiosity or the pursuit of knowledge. But the question is, How many researchers should take that path, and how many research budgets should be tied more closely to innovative achievements with specific purposes? Because the reality, especially (but not only) in Europe, is that the culture, motivation, and mindset in many large research institutions aren't much different from the mindset of impassioned bus ticket collectors. The interest in knowledge exists for its own sake. Goals are directed toward maximizing peer reputation. And finally—and perhaps a bit unkindly—scientific civil service and a passion for bus tickets both tend to be pleasant and low-risk activities. Which brings us back to the psychogram of breakthrough innovators.

Many conversations with HiPos end with them expressing, sometimes provocatively and sometimes in a friendly way, that they're tired of this zero-impact culture and want out. They want to be researching entrepreneurs and entrepreneurial researchers, because naturally they too know that knowledge has an impact above all in its application. On the other hand, this doesn't mean that most breakthrough innovators try to disguise their inner drive with aggressive improve-the-world-through-tech rhetoric or sell their idea as the one big thing that will save us all. When other people talk like that or refer to themselves as "tech evangelists" or something similar from the phrasebook of empty innovation-talk, it makes HiPos skeptical. They're afraid of being put on the same level as those techno-BS artists who may have a perfect sense of timing for surfing the latest wave of hype for a new technology, but (from the HiPos' perspective) don't recognize or understand other, better solutions that run counter to the trend. As a rule, this doesn't change their basic optimism about technology. And of course HiPos are also convinced that of all the possibilities, it's their solution that will advance medicine, information technology, materials science, energy or environmental technology, or whatever field they work in. But for the most part, that's just the motivational background context. Potential and actual breakthrough innovators are obsessed with solving their specific problem: The protein molecule that somehow has to pass through a cell wall. The static equilibrium of a giant wind turbine. The significantly higher energy density in a single battery cell.

NOT-SO-GREAT-MAN THEORY

In this thumbnail sketch of a psychogram for people with the potential to make innovative breakthroughs, there's naturally a good deal of sympathy and admiration. At the same time, we don't want to be understood as saying that the smart and obsessed people we've gotten to know from reading about them and having conversations with them are above average as pleasant colleagues and partners in cooperation. With all due caution about falling into clichés, we would go so far as to propose that in the inner life of the great inventors of past and present, there's a substantial portion of narcissism—and, in addition to technical excellence, outstanding skill for manipulation. Albert Einstein's presentation of himself in the role of the superstar among the natural scientists can be viewed with less sympathy than is usually the case. Among HiPos, we repeatedly find intellectual brilliance paired with cynicism and techniques of exercising power that are not at all in line with the principles of contemporary participatory leadership. Steve Jobs and Bill Gates were of course disruptive entrepreneurs. Not everyone who dealt with them will remember them for their remarkable fairness. As for Jeff Bezos, we remember the sentence of one of his ex-colleagues: "He just makes ordinary control freaks look like stoned hippies."[15]

It's also striking that it's almost always men in the spotlight when an innovative leap has been made. Who has heard of Gwynne Shotwell mentioned earlier as Elon Musk's congenial wingwoman at SpaceX? Have Jennifer Doudna and Emmanuelle Charpentier received the same attention outside the scientific community for the discovery and development of CRISPR-Cas9 as male colleagues would have, given the revolutionary importance of "genetic scissors" for innovations across the life sciences? And putting on our historian's hats, is there any reason other than gender bias why almost every child with an interest in technology knows Thomas Edison and Alexander Graham Bell, but has to google the names of Ada Lovelace (first algorithm), Grace Hopper (COBOL!), Barbara McClintock (genetic transposition), or Ann Tsukamoto (first stem cell isolation)?

At SPRIND, significantly fewer women apply than the agency would like. In its first years, the rate was under 10 percent. Now we are slowly improving, pushed by the women in our team, but by far not at the pace we wished. Here again, the causes seem to be complex. The balance of the conversations with male and female HiPos leaves the impression that this isn't the result of a conscious strategy to prevent female participation, but instead a mostly subconscious bias. There is increasing empirical evidence that women who found start-ups have a much more difficult time securing capital than men do, and there is also a double standard when it comes to risk-taking. Women are often punished for the same things that men are rewarded for. We'll go deeper into this in chapter 5 on financing, knowing full well that access to capital is an important problem but by no means the only one.

NINE GREAT INVENTIONS BY WOMEN YOU MIGHT HAVE NEVER HEARD OF

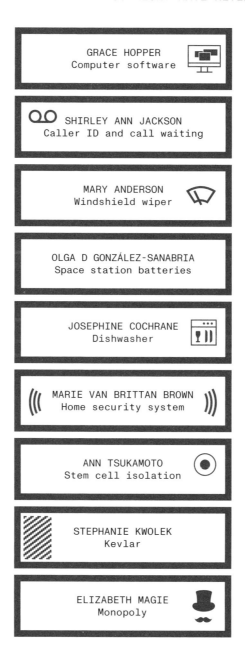

GRACE HOPPER
Computer software

SHIRLEY ANN JACKSON
Caller ID and call waiting

MARY ANDERSON
Windshield wiper

OLGA D GONZÁLEZ-SANABRIA
Space station batteries

JOSEPHINE COCHRANE
Dishwasher

MARIE VAN BRITTAN BROWN
Home security system

ANN TSUKAMOTO
Stem cell isolation

STEPHANIE KWOLEK
Kevlar

ELIZABETH MAGIE
Monopoly

Source: BBC, 2017

A lack of role models and network factors also seem to play an important role. Men with the ambition to be innovators—this too comes up repeatedly in conversation—more often have other men as friends who are already advancing innovation, and these friends then provide additional inspiration or support. That there is more than anecdotal evidence for this, and that it's not just a North American or European phenomenon, is suggested by an intriguing study entitled "More Women in Tech? Evidence from a Field Experiment Addressing Social Identity."[16] The two labor researchers, Lucía Del Carpio and Maria Guadalupe, from the French elite university INSEAD, conducted two field experiments in Peru and Mexico with potential applicants to a five-month software coding program offered to women from low-income backgrounds. When they corrected "the perception that women cannot succeed in technology by providing role models, information on returns and access to a female network, application rates double and the self-selection patterns change."

At least two conclusions can be drawn from Del Carpio and Guadalupe's findings. First, the study shows how challenging the cultural change is for creating parity and equal access for women in the sphere of innovation. Second, however, it proves that this is possible if the right information is provided using the right approach, if there is sufficient space for female role models, and if women's networks offer mutual support.

Unfortunately, data on the status of women in STEM, R&D, and tech are highly fragmented, incomplete, and difficult to compare between countries. This might be an indication of how large the deficits are in this area. As a proxy, however, we find it intriguing to look at the world map to see where women are well (or even very well) represented in STEM subjects at universities.

Many small states do quite well in this comparison, while the United States, most European countries, and even Asian countries with big ambitions for innovation (including China) do mediocre to poorly. To make matters worse, a relatively large number of women in STEM professions still work in the life sciences, chemistry, and material sciences.[17] Taken by itself, this is of course wonderful, but statistically it conceals a rather unwelcome development: the proportion of women in computer science in the United States has been cut in half since the mid-1980s, from around 40 percent to around 20 percent, and is now more or less stagnant at this level.[18] A key reason for this may be that since the late 1970s, home computers have consistently and successfully been advertised as gaming platforms and working tools for boys.[19] This marketing campaign following traditional gender stereotypes may have formed an important breeding ground for male-dominated geek culture, ultimately leading to today's Silicon Valley tech-bro culture with all its negative side effects. The bottom line is simple: you don't have to be a wizard with numbers to realize that if women aren't advancing innovation, we're wasting half the potential for innovation.

In order to raise women's potential as innovators, societies will have to make deep structural changes in their education systems, critically question gender identities, and create more objective narratives. It's also apparent that having more women involved in technological innovation can give technological development a different, more inclusive direction. This starts with simple safety features, such as seat belts, which have traditionally been optimized for male heights and weights by using "male" crash test dummies, and certainly doesn't end with how various fields of innovation are prioritized. Would treatments for menstrual pain have been researched more intensively over the past two hundred years if it was men who suffered menstrual cramping? Of course this question can't be answered ex post with empirical evidence, but the suspicion is highly plausible, especially if one adds up the large number of gender bias examples in tech that the British author and feminist activist Caroline Criado Perez collected in her eye-opening book, *Invisible Women: Exposing Data Bias in a World Designed for Men*.[20]

SHARE OF GRADUATES BY FIELD, FEMALE (%)

ECONOMY	YEAR	%		ECONOMY	YEAR	%
ARUBA	2016	40		ALGERIA	2018	58
AFGHANISTAN	2019	–		ECUADOR	2016	29
ANGOLA	2015	39		EGYPT, ARAB REP.	2016	37
ALBANIA	2018	47		ERITREA	2016	28
ANDORRA	2018	00		SPAIN	2017	30
UNITED ARAB EMIRATES	2017	41		ESTONIA	2017	39
ARGENTINA	2010	43		ETHIOPIA	2010	17
ARMENIA	2018	40		FINLAND	2017	27
AMERICAN SAMOA	2019	–		FIJI	2019	–
ANTIGUA AND BARBUDA	2012	33		FRANCE	2016	32
AUSTRALIA	2017	32		FAROE ISLANDS	2019	–
AUSTRIA	2016	26		MICRONESIA, FED. STS.	2019	–
AZERBAIJAN	2018	35		GABON	2019	–
BURUNDI	2017	18		UNITED KINGDOM	2016	38
BELGIUM	2017	56		GEORGIA	2019	39
BENIN	2015	55		GHANA	2018	20
BURKINA FASO	2018	21		GIBRALTAR	2010	00
BANGLADESH	2018	21		GUINEA	2019	–
BULGARIA	2017	37		GAMBIA, THE	2012	46
BAHRAIN	2018	41		GUINEA-BISSAU	2019	–
BAHAMAS, THE	2019	–		EQUATORIAL GUINEA	2019	–
BOSNIA AND HERZEGOVINA	2018	44		GREECE	2017	40
BELARUS	2018	27		GRENADA	2018	41
BELIZE	2015	42		GREENLAND	2019	–
BERMUDA	2018	20		GUATEMALA	2015	35
BOLIVIA	2019	–		GUAM	2019	–
BRAZIL	2017	37		GUYANA	2012	27
BARBADOS	2011	41		HONG KONG SAR, CHINA	2019	–
BRUNEI DARUSSALAM	2018	54		HONDURAS	2018	38
BHUTAN	2019	–		CROATIA	2017	39
BOTSWANA	2019	–		HAITI	2019	–
CENTRAL AFRICAN REPUBLIC	2019	–		HUNGARY	2017	32
CANADA	2016	31		INDONESIA	2018	37
SWITZERLAND	2017	22		ISLE OF MAN	2019	–
CHANNEL ISLANDS	2019	–		INDIA	2018	73
CHILE	2017	19		IRELAND	2016	29
CHINA	2019	–		IRAN, ISLAMIC REP.	2017	31
COTE D'IVOIRE	2019	–		IRAQ	2019	–
CAMEROON	2017	21		ICELAND	2012	35
CONGO, DEM. REP.	2016	25		ISRAEL	2019	–
CONGO, REP.	2017	21		ITALY	2016	40
COLOMBIA	2018	33		JAMAICA	2019	–
COMOROS	2019	–		JORDAN	2019	–
CABO VERDE	2018	42		JAPAN	2019	–
COSTA RICA	2018	32		KAZAKHSTAN	2019	32
CUBA	2016	40		KENYA	2016	31
CURACAO	2019	–		KYRGYZ REPUBLIC	2018	31
CAYMAN ISLANDS	2019	–		CAMBODIA	2015	17
CYPRUS	2017	38		KIRIBATI	2012	00
CZECH REPUBLIC	2017	36		ST. KITTS AND NEVIS	2019	–
GERMANY	2017	28		KOREA, REP.	2017	25
DJIBOUTI	2019	–		KUWAIT	2019	–
DOMINICA	2019	–		LAO PDR	2018	29
DENMARK	2017	34		LEBANON	2011	43
DOMINICAN REPUBLIC	2017	40		LIBERIA	2019	–

LIBYA	2019	–		RUSSIAN FEDERATION	2019	–
ST. LUCIA	2019	–		RWANDA	2018	36
LIECHTENSTEIN	2012	41		SAUDI ARABIA	2018	37
SRI LANKA	2018	41		SUDAN	2015	47
LESOTHO	2018	25		SENEGAL	2019	–
LITHUANIA	2017	30		SINGAPORE	2017	34
LUXEMBOURG	2016	28		SOLOMON ISLANDS	2019	–
LATVIA	2017	31		SIERRA LEONE	2019	–
MACAO SAR, CHINA	2018	25		EL SALVADOR	2018	23
ST. MARTIN (FRENCH PART)	2019	–		SAN MARINO	2018	36
MOROCCO	2017	45		SOMALIA	2019	–
MONACO	2018	00		SERBIA	2018	43
MOLDOVA	2018	31		SOUTH SUDAN	2019	–
MADAGASCAR	2018	31		SAO TOME AND PRINCIPE	2019	–
MALDIVES	2017	11		SURINAME	2019	–
MEXICO	2017	31		SLOVAK REPUBLIC	2017	35
MARSHALL ISLANDS	2019	–		SLOVENIA	2017	35
NORTH MACEDONIA	2017	48		SWEDEN	2017	35
MALI	2019	–		ESWATINI	2015	00
MALTA	2017	28		SINT MAARTEN (DUTCH PART)	2015	75
MYANMAR	2018	61		SEYCHELLES	2018	32
MONTENEGRO	2019	–		SYRIAN ARAB REPUBLIC	2016	50
MONGOLIA	2018	34		TURKS AND CAICOS ISLANDS	2019	–
NORTHERN MARIANA ISLANDS	2019	–		CHAD	2019	–
MOZAMBIQUE	2018	30		TOGO	2019	–
MAURITANIA	2017	29		THAILAND	2016	30
MAURITIUS	2017	36		TAJIKISTAN	2019	–
MALAWI	2019	–		TURKMENISTAN	2019	–
MALAYSIA	2018	34		TIMOR-LESTE	2019	–
NAMIBIA	2017	42		TONGA	2019	–
NEW CALEDONIA	2019	–		TRINIDAD AND TOBAGO	2019	–
NIGER	2018	18		TUNISIA	2018	55
NIGERIA	2019	–		TURKIYE	2014	35
NICARAGUA	2019	–		TUVALU	2019	–
NETHERLANDS	2017	29		TANZANIA	2019	–
NORWAY	2017	28		UGANDA	2019	–
NEPAL	2019	–		UKRAINE	2018	29
NAURU	2019	–		URUGUAY	2017	44
NEW ZEALAND	2017	35		UNITED STATES	2016	34
OMAN	2018	56		UZBEKISTAN	2018	25
PAKISTAN	2019	–		ST. VINCENT AND THE	2019	–
PANAMA	2016	43		GRENADINES	2019	–
PERU	2017	48		VENEZUELA, RB	2019	–
PHILIPPINES	2017	36		BRITISH VIRGIN ISLANDS	2019	–
PALAU	2019	–		VIRGIN ISLANDS (U.S.)	2019	–
PAPUA NEW GUINEA	2019	–		VIETNAM	2016	37
POLAND	2017	43		VANUATU	2019	–
PUERTO RICO	2016	39		SAMOA	2019	–
KOREA, DEM. PEOPLE'S REP.	2015	19		KOSOVO	2019	–
PORTUGAL	2017	38		YEMEN, REP.	2019	–
PARAGUAY	2019	–		SOUTH AFRICA	2017	43
WEST BANK AND GAZA	2018	44		ZAMBIA	2019	–
FRENCH POLYNESIA	2019	–		ZIMBABWE	2015	29
QATAR	2018	48				
ROMANIA	2016	41				

Source: Gender Data Portal – The World Bank, 2022

It likewise seems plausible to assume that having more women involved in innovation would also accelerate a paradigm shift toward "technological progress" in the humanistic sense. Again, there's no proof ex ante, but there are strong indications. Women select into social professions more than men, partly because they hope to increase others' happiness. Why should this be any different in technological development? Perhaps it will help increase the number of women making innovative leaps at least a little bit to remind readers of Marie Curie, who researched jointly with her husband Pierre and shared the Nobel Prize in Physics with him. In the iconic photo of the 1927 Solvay Conference, Curie, the only woman in attendance, looks confidently into the camera. Marie Curie broke another record: she was the first person ever to be awarded a second Nobel Prize, for chemistry.

We'll close this chapter with another note that runs the risk of repeating a cliché, but in our perception, it describes reality better and somewhat less humorously than *The Big Bang Theory* TV series. When non-nerds get access to conversations that nerds have among themselves, they're temporarily struck with fear. Potential breakthrough innovators tend to bubble with optimism as they discuss technological possibilities, scenarios, and goals that would have difficulty gaining the support of even 5 percent of the general public in a representative survey. Among nerds, there have always been absolute majorities in favor of nuclear power, with a particular soft spot for nuclear fusion. They're often not afraid of machine super-intelligence—an AI superior to human beings in all respects—but instead look forward to its arrival and enthusiastically debate which theoretical technological hurdles still have to be overcome to achieve it. They often hope that soon people will be able to live a thousand years, if not forever. And if this isn't possible in a purely biological sense, then they're just as interested in a melding of man (or woman) and machine. Because only this will allow human beings, or our cyborg descendants, to travel to exoplanets in neighboring galaxies. There the remainder of our biological humanity can continue to live as human–machine hybrids when our planet becomes uninhabitable, whether as a consequence of human stupidity, asteroid impact, or a burst of cosmic radiation that human beings are powerless to stop.

There is disagreement between the two authors of this book as to how likely and desirable these scenarios are and whether they are likely to occur only a few generations hence, or if at all, then in a completely unpredictable future many thousands of years after we, like most of us, are buried in the earth of this planet, and not like the bones of Elon Musk on Mars. On the other hand, we agree that HiPos need more support and better working conditions in their areas of activity than they have previously had, among other things so they can follow the example of Akira Endos and develop more effective medications.

How do we improve the context and working conditions for innovative leaps, and what roles should the state and the market play in this? The next two chapters are devoted to these basic questions.

Akira Endo, incidentally, never had any financial stake in the hundreds of billions of dollars that pharmaceutical companies have earned with statins since 1987. The Japanese breakthrough innovator never said a word about it in public. Like many nerds, the fungus nerd wasn't really interested in money. Mold was more important to him.

THE ENTRE-PRENEURIAL STATE

HOW CAN FRESH POLICY THINKING FOSTER INNOVATION?

RUN, ROBOT, RUN

Running Man has mastered all the obstacles on the course. He first got into a car and drove to the disaster site. Then the robot climbed over debris and cleared away heavy concrete rubble. Running Man opened a door, ran up a flight of stairs, and broke through a concrete wall. And finally, the half autonomous, half remote-controlled machine closed a valve on a leaky pipe, attached a hose to a fire hydrant, and turned on the water. At the finish line, he raises his steel arms into the air. The spectators in the stands of the Fairplex Arena in Pomona, California, cheer for the humanoid robot from the Institute for Human and Machine Cognition. Running Man then performs a short, awkward victory dance. The robotics fans laugh, cheer even louder, and clap rhythmically. A little nerd humor comes as a relief for everyone at the 2015 Robotics Challenge of the Defense Advanced Research Projects Agency (DARPA) after three years of hard work on a major technological challenge.[1]

What is easy for humans is difficult for robots, and vice versa. That's an old rule of robotics. Running Man needed around fifty minutes for the eight tasks in the disaster scenario devised by the US Department of Defense's research and development agency. A moderately trained firefighter with a battery-powered impact drill in a backpack could have done the job ten times as fast. But the Grand Robotics Challenge wasn't a competition between human beings and machines; it was a search for answers to the question, Can robots be reliably deployed in places where humans can't go? One possible scenario would be an accident in a nuclear power plant on the brink of meltdown. That is precisely how the exercise in Pomona was set up.

In the competition between twenty-three robots from North America, Asia, and Europe, Running Man ultimately came in second to HUBO, the entry from a team of young students from the Korea

115

THE ENTREPRENEURIAL STATE –
HOW CAN FRESH POLICY THINKING FOSTER INNOVATION?

WHAT'S EASY
FOR HUMANS
IS DIFFICULT
FOR ROBOTS,
AND VICE
VERSA.
THAT'S AN
OLD RULE OF
ROBOTICS.

Advanced Institute of Science and Technology, so the Korean team pocketed DARPA's $2 million in prize money. But at the end of the last day of competition, nobody seemed unhappy. The robots had done their jobs much better than the experts had expected at the start of the challenge three years previously. DARPA had deliberately made the obstacle course difficult. In the end, the semiautonomous machines' learning curve on robot-hostile terrain was steeper than people had expected.

Marc Raibert, chairman at Boston Dynamics, compared the atmosphere in the stands and in the workshops in the team camp with a "Woodstock for robots."[2] Participants spoke of a "spirit like in the Olympic village." Once again, the world's largest governmental innovation agency had shown that promoting innovation with tax money can be both fun and technologically effective. Over the long term, the public attention it can generate will benefit the competing teams and everyone else working in the same field of research. Another by-product was open-source software that robotics developers around the world can access free of charge as well as the open sharing of general design knowledge for everyone hoping to advance the field of robotics. That is exactly what people had hoped to achieve.

Behind the challenge's specific question concerning how robots might be deployed in a disaster, there's an overarching philosophy of state support for innovation. The (human) head of the Robotics Challenge, DARPA program manager Gill Pratt, sums up this philosophy as follows: "What we do is we wait for technology to be almost ready for something big to happen, and then we add a focused effort to catalyze the something. It doesn't mean that we take it all the way into a real system that's deployed or to the marketplace. We rely on the commercial sector to do that. But we provide the impetus, the extra push the technology needs in order to do that."[3]

Since 2015, semiautonomous robots have significantly improved their skills and dexterity in harsh, nonstandardized environments. We can watch their progress in thousands of videos on YouTube. Robotic arms cook food and empty dishwashers, collaborative robots ("co-bots") relieve human colleagues of difficult physical labor, and semiautonomous rovers drill for water on Mars. And of course, robot dogs also do somersaults while robotics technicians take them for walks downtown. Scenes like this get the most clicks. Projects with rescue robots following directly from the Robotics Challenge have also made significant progress since 2015. That's impressive, but compared to many of DARPA's innovation initiatives since it was founded over sixty years ago, the provisional results are fairly modest.

100 GENIUSES AND A TRAVEL AGENCY

As for many new government high-risk, high-reward innovation agencies and programs like ARIA in the UK or the Moonshot in Japan, or ARPA-H and ARPA-E in the United States, DARPA is an important role model for the Federal Agency for Disruptive Innovation—its tradition of open innovation competitions, and of course its serving as the inspiration for SPRIND's antiviral drug platform challenge (in which we also included the element of "forced cooperation" as it was used in DARPA's mRNA project). In this sense, our perspective on the Defense Advanced Research Projects Agency may be subject to affinity bias.

And we're well aware that the history of DARPA has seen its share of minor scandals and justified concerns over whether individual decision-makers might have engaged in favoritism. In con-

117

THE ENTREPRENEURIAL STATE –
HOW CAN FRESH POLICY THINKING FOSTER INNOVATION?

fidential conversations, European participants in DARPA programs have admitted that despite its advertised collaborative principles, micromanagement isn't an entirely alien concept at DARPA. And of course, the organization's—desired and necessary—autonomy, which constitutes the core factor in its success, also raises difficult ethical questions. For example, is it morally justifiable for a government innovation agency with a military background to conduct research on neurotechnologies in which computer chips are implanted in soldiers' brains to turn them into invincible super-soldiers? According to critics, DARPA's almost limitless ambition for innovation leads the organization to reflexively brush aside urgent ethical discussions.[4] This isn't only true of military research, by the way. Radical biotechnological innovations always raise numerous ethical questions. All of that is correct and worth thinking about.

But American critics shouldn't underestimate how attractive the DARPA model is, including for ethically scrupulous breakthrough innovators, researchers, and research sponsors around the world, who are subject to bureaucratic control freaks in their research and development systems. Which is to say, precisely those researchers whose innovations are repeatedly delayed or thwarted in accordance with the happiness-maximizing principles of the SDG agenda discussed in chapter 2. That's why we're going to take a closer look at DARPA's history and working methods.

The US innovation agency is a child of the Cold War. President Dwight D. Eisenhower founded it in 1958 as a direct response to the so-called Sputnik shock.[5] In 1957, Soviet scientists had succeeded in launching the first communications satellite into low Earth orbit using a modified R7 intercontinental ballistic missile. The United States, normally self-confident in its position as a high-tech nation, felt caught off guard, uncertain, and challenged. DARPA's mandate was accordingly to restore America's undisputed high-tech sovereignty, including for military reasons during the Cold War, but also with the involvement of all useful partners from science, industry, and other public sectors without a direct relationship to the Pentagon. At the beginning—and later intermittently—the agency didn't have the "D" for "defense" in its name.

In the first year of its existence, a turf war raged between ARPA and NASA, which was also founded in 1958. ARPA's founding director, a former General Electric manager and space travel enthusiast, resigned after being defeated in the battle for responsibility for research on outer space. There may not be such a thing as a government institution that doesn't experience birth pangs without extraterrestrial help. But in the 1960s, ARPA quickly developed into an organization that might even impress another intelligent species. It was supposed to advance "high risk," "high gain," "far out" research—goals and principles for which governmental research funding and project sponsorship were unfamiliar, even at the time. This meant that the innovation agency needed an unbureaucratic structure and culture, or a radical innovation in administration itself. In this case, removing the red tape from bureaucracy actually succeeded.

Today, DARPA has an annual budget of around $4 billion.[6] This money isn't managed, but rather invested under the direction of around 220 staff members in a small, agile organization with an extremely flat hierarchy and an unusual distribution of decision-making powers. There's a small board of directors and staff who have to approve new research programs, define guidelines, and ensure a balanced "investment portfolio." A second flat organizational level consists of the technical offices. This level bundles knowledge and coordinates the organization's critical personnel, the circa 100 program managers. These program managers are usually well-connected innovators in their own right from science labs or tech

WITHOUT (D)ARPA, IT'S UNLIKELY THE APOLLO MISSION WOULD HAVE MADE IT TO THE MOON.

119

THE ENTREPRENEURIAL STATE –
HOW CAN FRESH POLICY THINKING FOSTER INNOVATION?

companies. They move to DARPA for three to five years in order to "push their discipline to its own limits," as the organization sees their mission. To do this, they're constantly traveling to talk to inventors, research managers, and founders of deep-tech start-ups. "100 geniuses connected by a travel agent," as a former director put it, became a catchphrase for the agency and among those at the Pentagon who allocate its budget.[7] The best program managers, for their part, are acknowledged as "freewheeling zealots."[8] In an ideal situation, they add "A and B to get Z." In these self-descriptions, there may be a touch of techno-vanity and the American talent for self-marketing. Depending on the context, it can seem interesting and humorous, or boastful and self-important. But there's little room for interpretation in answering the question of whether or not these freewheeling fanatics who add A and B together to get Z have been responsible for an impressive number of military and civilian breakthrough innovations. They have.

Without (D)ARPA, it's unlikely the Apollo mission would have made it to the Moon because it developed the engines for the Saturn rocket. Radar and sensor technology owe a whole series of innovative leaps to the tech geniuses and their travel agents, and so does the radar-evading stealth technology used in the B-2 and F-35. Without DARPA, there would probably be no satellite-based GPS navigation in our cell phones and cars today. The computer mouse and graphical user interface were the offspring of ARPA projects, as was probably the greatest innovative leap of our time: ARPANET, the forerunner of the internet, which initially was intended to serve as a decentralized and therefore particularly resilient military communication network in the event of a nuclear war. On behalf of ARPA, the programming visionary Vinton Cerf developed the TCP/IP network protocol, which made it possible for the internet to network every person and computer in the world without being intended for that purpose. DARPA continued to support research on machine learning when other institutions choked off funding for artificial intelligence, including what was called at the time the Speech Understanding Research program (SUR), the forebearer for Siri and Google Assistant. It helped the discipline survive the infamous "AI winters" before machine learning really became useful a decade ago thanks to the wealth of data from the internet and the computing power of modern microchips, also funded by DARPA. Submarines and drones have always belonged to DARPA's favorite projects, as have robotics, bionic limbs, and a liquid to foam substance that stops bleeding instantly. More recently, IT security and defense against cyberattacks have gained increasing importance in research, which also benefits civilian IT users. Perhaps the next greatest acceleration of an innovative leap after ARPANET came from a predecessor to the robotics competition, DARPA's Grand Challenge for autonomous driving in 2004.[9]

As with the rescue robots, DARPA invited the public to an extravaganza in California, at the time on a restricted military site in the Mojave Desert with a 150-mile route marked out on uneven dirt roads. A million dollars in prize money awaited the team whose autonomous vehicle traveled the entire route without human intervention. The winning team made it just 7.4 miles that year, and American tech journalists made fun of "DARPA's debacle in the desert." Historically, this would prove to be a gross misjudgment. The next year, five vehicles successfully completed a 132-mile course. First to the finish line was the vehicle from Stanford University, the development of which had been led by the computer scientist Sebastian Thrun. The noticeable progress in the 2005 competition (prize money: $2 million) laid the groundwork for Google's decision to massively invest in self-driving cars. Shortly thereafter, Thrun became head of development for autonomous vehicles at Google, which in turn sent the signal to competitors that the internet's dominant search

121

THE ENTREPRENEURIAL STATE –
HOW CAN FRESH POLICY THINKING FOSTER INNOVATION?

engine was serious about the technology. Tesla, Uber, the traditional car manufacturers, several start-ups, and later the Chinese Big Tech companies entered the race for self-driving cars as well. Two things are now foreseeable. First, at some point in the coming years we'll take a seat in cars that largely drive themselves. Second, technology historians will once again see the Grand Challenge 2004 as one more success for DARPA in giving technology a powerful boost through the intelligent use of tax funds.

However, there's a third notable aspect demonstrated by the successes of DARPA in general and the Grand Challenge for autonomous driving in particular. There's a persistent myth that in the United States, the market, venture capital, and start-up culture are the main drivers of radical innovation. This perception is both widespread outside the United States and an integral part of the United States' self-image. It's time to expose this myth to tough questioning. In the United States, with its supposedly radical market orientation, it's the government that's making massive targeted investments in innovation in the style of Schumpeter with ambitious goals—and afterward taking an exacting look at the result, at the "return on investment." The rise of Silicon Valley to become the most important hotspot for innovation in the world wouldn't have been possible without massive state investment in the computer chip industry and regular government orders.[10] Tesla, by far the most valuable automaker in the world today, would likely have ceased to exist in 2009 if the American government hadn't saved it using taxpayer funds.[11] In the long term, those were smart investments. In contrast to many other countries that often take pride in a less market oriented, more government driven approach to the advancement of society, including Japan, South Korea, France, and Germany, the United States is an "entrepreneurial state," at least in the field of breakthrough innovations. This term wasn't invented by DARPA public relations staff. It was coined in 2013 by an Italian

American economist who researches and teaches at University College London and is touted as a future winner of the Nobel Prize in Economics.

THE ENTREPRENEURIAL STATE

Scholars in the field of innovation studies with an interest in the intersection of economics, technology, and socioecological development were familiar with the name Mariana Mazzucato even before 2013. She had made a name for herself in the academic community as an astute analyst of Schumpeter's theory of innovation in the age of PCs, the internet, biotech, and Big Pharma. Outside the economic mainstream, Mazzucato created a hybrid between Schumpeter's ideas and post-Keynesian thought—that is, the belief that the state should manage business cycles through massive investment and thereby address society's major challenges. Like Romeo and Juliet, market meets state. Joseph Schumpeter and John Maynard Keynes, whose mutual disdain and hostility was legendary, would probably be spinning in their graves if they could read Mazzucato's 2013 book *The Entrepreneurial State: Debunking Public vs. Private Sector Myths.*[12] Or perhaps the two adversaries, still grumbling at each other, would have learned to get along. Mazzucato intelligently addresses the traditional conflict between the "too much state is always harmful" and "the market always fails" schools of thought, and if she doesn't fully resolve the conflict, she at least succeeds in elevating it to a higher level. She thus makes possible a compromise between Schumpeter fans with their market fixation and neo-Keynesians who believe in the state.

She begins her analysis by noting that the celebrated (and self-congratulatory) entrepreneurial heroes of innovation should

123

THE ENTREPRENEURIAL STATE –
HOW CAN FRESH POLICY THINKING FOSTER INNOVATION?

occasionally consider the fundamental innovations out of which they construct their "disruptive" products—and just who funded and created these basic innovations. Mazzucato enumerates a seemingly endless series of examples, from IT to materials science, from nano- and environmental technology to the life sciences, where public institutions and government labs, state-funded basic research, and the major governmental innovation agencies created a radically new technology that entrepreneurs only later brought into broad use. The most striking example of all is shown in the accompanying figure, the technological genesis of the iPhone.[13]

Mariana Mazzucato empirically refutes the widespread perception that governmental research and development institutions are bureaucratic behemoths that devour vast sums of money while adding little economic value, at least for the United States. This historical analysis is in turn the source of her plausible and hopeful vision of an entrepreneurial state. In an entrepreneurial state, there isn't a constant dispute about whether more money should be spent on research, or whether it will only be absorbed by inefficient research bureaucracies. Rather, the consensus is that the government and innovative companies take risks together, coordinating more often and more effectively, in order to accelerate innovation in a mission-focused way and target the challenges facing society in medicine, ecology, and access to prosperity. In this model, the entrepreneurial state is supposed to enjoy a much greater share of the profits that new, publicly developed technologies, such as the internet, touch displays, GPS, speech recognition, and MP3 audio compression, generate for the shareholders of Apple, Google, and other tech firms, including through ownership of shares in the companies.

In our perception, Mariana Mazzucato's tone occasionally takes the Keynesians' side too strongly, including in her new book *Mission*

Economy: A Moonshot Guide to Changing Capitalism.[14] She mostly ignores the fact that society broadly and directly benefits from the use of innovative products (the so-called consumer surplus), and that the state already earns a considerable amount from successful products through tax revenues. Sometimes you get the impression that Mazzucato sees the state and entrepreneurs as opposing teams on the field of innovation. But in essence, her books and essays get to the heart of something that's often subtext in the great debates in innovation research but that's seldom clearly and explicitly expressed: before any market exists, the state must act as a risk-taker. Working together with companies, the state has to shape new markets that arise through innovation so that added value is created for society. The state can and should set innovation agendas and mission deadlines. Businesses and their well-paid agents need to acknowledge this role of government more clearly. At the same time, they need to optimize their innovation strategy less around the goal of tapping as much funding or subsidized technology as possible and then, if they're successful, minimizing their tax burden with legal but questionable corporate tax avoidance techniques.

RED TAPE, RED FLAGS, RED CARPET

How can countries be more radically innovative than they are today? What role should government play? And what lessons can be drawn from DARPA's success? We see three important levers for action:

125

THE ENTREPRENEURIAL STATE –
HOW CAN FRESH POLICY THINKING FOSTER INNOVATION?

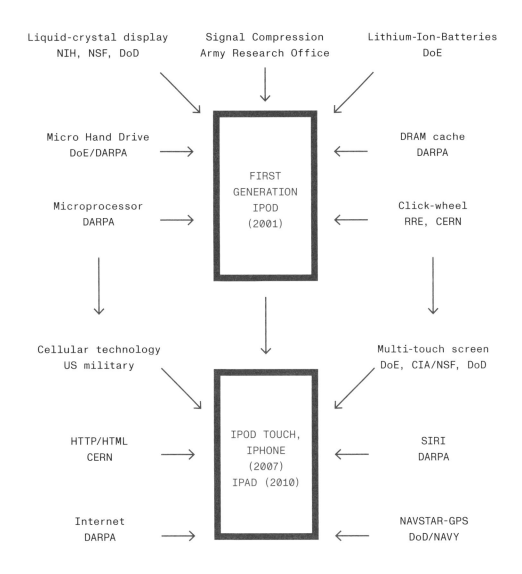

NIH: National Institutes of Health
NSF: National Science Foundation
DoD: Department of Defense
DoE: Department of Energy
DARPA: Defense Advanced Research Projects Agency
RRE: Royal Radar Establishment
CERN: European Organization for Nuclear Research
CIA: Central Intelligence Agency

Adaptation: Mariana Mazzucato, *The Entrepreneurial State*

$\rightarrow$ 1.

Radically Removing Red Tape from State Funding for Innovation
Obviously, not every aspect of research and development funding
can work like DARPA, where intrinsically driven nerds in the role
of program managers quickly cut large checks for other HiPos in the
role of breakthrough innovators. It goes without saying that a research
funding system that distributes billions of dollars in tax funds needs
sufficient mechanisms for assessment and monitoring. But it is crucial
to ask how much is sufficient. It seems at least as clear to us that the
jungle of guidelines and eligibility criteria are overseen by a small
army's worth of national civil servants and bureaucrats. In making
funding decisions, they're legally bound by a tangled mess of regulatory safeguards. The result is the opposite of what sensible funding
regulation actually hopes to achieve: both procedural propriety and
fairness in awarding funding as well its efficient use, ultimately resulting in numerous innovations. At its core, it's about the old question
of control and effectiveness. Or to question a piece of conventional
wisdom, in statistical terms, is it really true that trust is good, but ver-
ification is better? Or does an excessive need for supervision, which
already controls the decision-making process for innovation funding,
hobbles our progress?

Part of bureaucracy's positive contribution is its reliability.[15] Ideally, this is paired with equal treatment and transparency. A willingness to make decisions under conditions of uncertainty and to make
and admit to mistakes is not in the nature of most bureaucrats. This
is all understandable to a certain extent, and we too think bureau-
crat-bashing, whether at the state, federal, or international level, is
of little use. That being said, research funding organizations all over
the world could use a good dose of DARPA culture. Innovation in
general, and innovative leaps in particular, constitute a game where

127

THE ENTREPRENEURIAL STATE –
HOW CAN FRESH POLICY THINKING FOSTER INNOVATION?

the right people need the right support at the right time. Time pressure caused by international competitors is almost always a factor. Unfortunately, the legitimate bureaucratic desire to follow procedure correctly to prevent cronyism and eliminate the risk of waste from the outset too often lands jam-side down.

The traditional regulatory insanity attracts the wrong people, namely those who are skilled at navigating bureaucracy to secure subsidy funding instead of those with a deep interest in deep tech. Corporations, large research institutions, and organizations that sponsor projects have built up departments and expertise over the decades to overcome the hurdles that stand between them and subsidy funding. When we talk to medium-sized innovators and deep-tech start-ups from the United States or Europe, we often hear the comment, "If I have to fill out hundreds of pages of application forms, I don't have any time left for developmental work." Conversely, when we read the interim and final reports of funded science projects, we notice that there's always a carefully placed teaser for the need for follow-up financing, even if it's only to check out "how much milk is left in the cow," to quote a research manager who (naturally) wishes to remain anonymous. The award processes are slow, extremely detailed, and, in their regulatory complexity, often inconsistent. In the end, despite all the good intentions, the result is the highest degree of waste: countless billions invested and very little innovation that ever reaches citizens. The bottom line at this point is that research and innovation don't follow the logic of a bureaucratic request for proposals. Innovators shouldn't need to know in advance how everything works. They cannot, by definition, as they explore new territory. If they do, they are not innovative. They have to be allowed to fail. Research funding has to be available, not only to large institutions that can afford the bureaucratic effort and the snail's pace of approval.

THE   T

T

INSA

RADI-
IONAL

NITY

To change this, we need both a clean break from existing legal frameworks and a cultural transformation. State-based subsidies and public procurement law need to be radically streamlined. In terms of innovation policy, for example, it makes no sense to prohibit research funding from supporting multiple competing teams who are working on the same problem simultaneously. To bureaucrats it might seem wasteful to let several people do the same thing. In exploring new territory, it's not. The heart of the innovation game is competition. Betting on the race is smarter than betting on a horse.

We also need to fundamentally rethink how we treat intellectual property that was created by brilliant minds using taxpayers' money. This applies above all to spin-offs from public research institutions. Research bureaucrats often take a very short-term view of returns on public investment. Of course, the state and its taxpayers have the right to share in an economic success made possible with state-funded intellectual property (IP). Unfortunately, many public funding institutions have failed to understand that playing the long game is the only way to win, and if they can't do that, it's best to stay out of it. It's better to be long-term greedy than short-term greedy. What does that mean in concrete terms?

When a team at a government-funded research lab makes a discovery that has spin-off potential, the team should be allowed to use all the IP without facing legal pitfalls.[16] In return, all they have to do is grant a small percentage of nonvoting shares in the company to the (governmental) scientific parent organization. This significantly increases the chances of obtaining growth capital and economic success, which in turn is easy to explain in economic terms. Capital is extremely expensive for start-ups. The more cash they have to pay for IP at their founding, the more shares they have to cede directly to investors. That by itself is already an inauspicious start. On top of that, the incentives for venture capitalists sink if start-ups first have

131

THE ENTREPRENEURIAL STATE –
HOW CAN FRESH POLICY THINKING FOSTER INNOVATION?

to use their precious capital to pay for expensive licenses that bureaucrats demand upfront because, in their logic, it's only fair. Then, if the bureaucrats also want to stick their fingers into business development, venture capitalists' interest in investing goes cold. That's no way to create the next decacorn. Too bad for the founders, a bad break for innovation—but also tough luck for the taxpayers who prefinanced the IP.

The privately run research organizations that have seen major spin-off successes, such as Harvard, Stanford, and MIT, on the other hand, understand that from an economic point of view, having a share in a spin-off is like having a lottery ticket that you've already paid for anyway, and at the same time their approach has a much better chance of winning. The more tickets an organization buys, so to speak, the higher the chance of winning the jackpot. So the obstacles for spin-offs need to be low, especially for assigning IP to the founding teams. In the short term, it's an act of generosity. In the long term, it might even come to be seen as greedy, as the enormous returns enjoyed by the private spin-off hatcheries show. Playing a sufficient number of lottery tickets leads to big profits. An entrepreneurial state can also be greedy in the long term. But if it only eyes the short-term profits that make life difficult for founders from the start, it shows that the state has a complete lack of entrepreneurial skill.

Meanwhile in Europe, bureaucratic matters are even more complicated than US scientists and innovators—at least those not associated with DARPA, but with traditional funding organizations like the National Institutes of Health—might imagine. One of the reasons is a strained relationship between national and European innovation funding. National funding in Europe isn't allowed to discriminate against innovators in other European countries.[17] That sounds good and fair, but it hobbles national funding in many ways, which can't be in the best interest of Europe as a whole. It would be logically consistent to award innovation funding exclusively on a European level, but that wouldn't promote a diverse and fertile innovation landscape in all EU member states. This structural error in the subsidy funding system needs to be eliminated through legislative action. But that strikes us as almost an easier task compared to the cultural and organizational issues.

Reducing red tape as a step toward becoming an entrepreneurial state can only succeed if there's a real innovation process in the institutions and in the minds of the funding bureaucracy. As of today, wasted money isn't regarded as a problem as long as the waste is sufficiently justified and settled in accordance with rules and regulations and thus reinterpreted as a productive use of funds. Because then nobody in the bureaucracy can be said to have made a mistake. The operation was a success, but the patient died. In a system where no mistakes are allowed, no one makes mistakes by definition, because all eventualities have previously been addressed by all necessary safeguards. Even if it sounds trite, innovation is based on the courage to take risks. It depends on the ability to recognize and name what's working well and what isn't. The goal isn't to make mistakes, but we have to learn to let them happen. They'll be made anyway, and only a culture that accepts mistakes makes learning effects possible. And after all, the stunning success of DARPA cranking out breakthrough innovations every few years is based on the much higher number of failures it has produced.

133

THE ENTREPRENEURIAL STATE –
HOW CAN FRESH POLICY THINKING FOSTER INNOVATION?

2.

Measure Innovation, Reward Success,
Red Flag Pseudosuccess

NO SCIENTIST SHOULD HAVE TO JUSTIFY HER-/HIMSELF TO COLLEAGUES FOR HAVING THE GOAL OF CREATING KNOWLEDGE WITH A PROSPECT OF PRACTICAL APPLICATION.

"If you can't measure it, you can't manage it." This quote is among the best-known legacies of the economist and management visionary Peter Drucker. Ironically, it seems Drucker has never said it, as his vision of measuring was a bit more nuanced.[18] Nonetheless, Drucker's (more nuanced) thinking on measuring provides a critical guideline for how we can organize innovation processes at the intersection of research, government, and the market so that these processes don't get bogged down in bureaucratic trivia but instead promote and support innovation in a results-oriented way. The two crucial questions here are, of course, what and how do we measure.

Innovation is especially difficult to quantify in the early phase of the process because there are usually no firm indicators of whether an idea will turn into a successful product, a more efficient process, or even a radically better solution with consequences in many different areas. Unfortunately, this has led to two counterproductive developments in the allocation and supervision of research and development support in the current funding system. First, especially in the area of basic research, you can frequently sense reluctance on the part of many researchers to face a real evaluation: an evaluation process that seriously asks what the goal of the research project actually is and how this goal, assuming it is achieved, would add scientific, social, or economic value. When people who formulate research policy ask this question, it immediately raises the suspicion that they don't have proper respect for scientific freedom. But did Wilhelm von Humboldt really mean for his intellectual descendants to avoid sullying

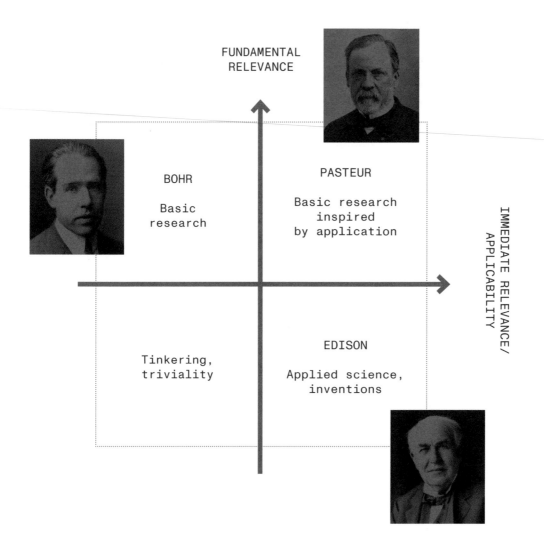

FUNDAMENTAL
RELEVANCE

IMMEDIATE RELEVANCE/
APPLICABILITY

BOHR

Basic
research

PASTEUR

Basic research
inspired
by application

Tinkering,
triviality

EDISON

Applied science,
inventions

135

THE ENTREPRENEURIAL STATE —
HOW CAN FRESH POLICY THINKING FOSTER INNOVATION?

their pure research with economic value and practical application?[19] We think Humboldt might see this as a distortion of the Humboldtian model of the university. We aren't the only ones with the impression that the overly strict separation between basic research and application-oriented research seems increasingly too one-dimensional—and it's also not a phenomenon limited to Europe. In the United States as well, there's an increasing call for researchers in basic science to define themselves less in contrast to application-oriented researchers; their model should instead be Louis Pasteur. In 1997, the American political scientist Donald Stokes introduced the model of "Pasteur's quadrant" to the discussion.[20] Thomas Edison was only concerned with application, Niels Bohr only with theory, but the greatest advance came through the microbiologist Louis Pasteur, who introduced both the theoretical knowledge and the practical application of vaccinations.

The world could certainly use more Edisons and Bohrs today. But no scientists should have to justify themselves to colleagues for having the goal of creating knowledge with a prospect of practical application. Precisely that is often the case in today's academic culture, however.

The reflexive invocation of "academic freedom" as soon as anyone brings up measurable criteria for supporting research and scholarship often sounds in our ears like researchers looking for an excuse. And to us, it repeatedly comes across as a lack of understanding of the role of science as an activity financed by society to lead the way to social progress and prosperity. When pressed, a common evasive maneuver is to point to seemingly hard data, such as citation counts or an H-index. But the number of high impact publications, which may be a more meaningful indicator of the actual contribution to innovation, gets swept under the table if it's not as flattering. And it strikes us as almost grotesque when the share of research funds drawn from a particular funding source is taken as a key measure of research success.

When it comes to documenting successful innovation based on research funding, probably the most reliable data are the number of start-ups spun off from a research organization; the number of good jobs created; the sales, profits, and company market values attained by the start-ups; the returns to the state and society through corporate taxes paid; or in the case of a spin-off in environmental technology, the tons of carbon dioxide prevented from entering the atmosphere. It doesn't take statistical sorcery to collect these data and present them clearly. Successful partnerships between research institutions and companies can also be measured effectively, for example, by sales of new applications or products based on registered patents. These numbers can also be put to effective use to direct subsidies more efficiently. A reasonable assumption might be that a careful accounting of impact-oriented indicators would find that the inputs into research funding are out of all proportion to the innovative output of the major traditional research organizations.

And newer organizations might actually be doing better than many critics from the science establishment claim. A good example of this is ARPA-Energy, conceived under President George W. Bush and then made a reality by the Obama administration. Anna Goldstein, who researches energy policy at the University of Massachusetts Amherst, compiled the numbers and compared them to the usual output of research funding. On a scientific level, ARPA-E-funded researchers produce a significantly higher number not only of high-impact research papers but also—and more importantly—patents.[21] The agency's own published figures are just as impressive. From its inception in 2009 through April 2022, ARPA-E has awarded approximately 3 billion dollars in R&D funding to more than 1,300 innovative energy projects. This has led to 900 patents and technologies that have been licensed

137

THE ENTREPRENEURIAL STATE –
HOW CAN FRESH POLICY THINKING FOSTER INNOVATION?

300 times. Added to that, some 190 projects have raised more than 10 billion dollars in private-sector follow-on-money, which in turn has enabled the creation of 130 start-ups. However, the most interesting performance indicator here is that among the ARPA-E cohort of start-ups, twenty-five founding teams have made a successful exit via mergers and acquisitions or initial public offerings. The total value of the companies amounted to almost 22 billion dollars, or more than seven times the amount of taxpayer funds used.[22]

It's still too early to tell whether the products' and technologies' level of invention and the ARPA-E start-ups' long-term commercial success are greater than those of start-ups that go without public funding. This naturally requires long-term evaluation, as deep technologies take a long time to build speed. However, it seems particularly important to us to measure this indicator reliably in order to optimize future funding in terms of impact. According to energy policy expert Anna Goldstein in an interview with Nature, there's already clear evidence in favor of ARPA-E: "The answer is yes, the ARPA model works."[23]

For transitioning to more effective management of innovation transfer, it could help to collect somewhat softer leading indicators that suggest that a scientific organization is at least increasingly committed to the ambition to transfer new knowledge. To do this, it would be important to know how many young researchers have been trained in entrepreneurial thinking. Logically, directing resources based on measurable evidence would also mean that pseudosuccess alone doesn't justify continued support. Agile, output-oriented innovation funding from an entrepreneurial state needs the courage to end projects earlier.

This closes the circle of cultural transformation toward unbureaucratic state funding for research and development along with rigorous evaluation, true to the mantra of the expert commission for

DARPA HAS A
COMPETITIVE
ADVANTAGE
THAT NO
OTHER
RESEARCH
FUNDING INS-
TITUTION IN
THE WORLD
CAN EVEN
COME CLOSE
TO OFFERING:
THE PURCHA-
SING POWER
OF THE US
MILITARY.

innovation: measure, measure, measure! Measuring success also means measuring failure. Failed research projects don't initially constitute a mistake, or more precisely, a faulty decision on the part of those who approved funding. They only become a mistake when they continue to be supported because no one wants to admit that an investment didn't pay off. Behavioral economists like Daniel Kahneman and Amos Tversky have shown in many experiments how difficult it is for us to accept sunk costs—costs that can't be undone or covered by revenue.[24] We tend to throw more and more money into bad investments because we can't admit to having thrown away money. Bureaucracies especially suffer from this bias particularly because the need to justify every decision is so firmly anchored in the bureaucratic system. By their nature, investments that are intended to enable technological breakthroughs have a chance of failing. Those who fail to overcome sunk cost bias will have no money left to invest in the next breakthrough. Ken Gabriel, former DARPA director and current codirector of Wellcome Leap, says, "It is acceptable to fail. What's not acceptable is if you don't know whether you fail."[25]

Not measuring success and failure is clearly a red flag for innovative funding. But to avoid any misunderstanding, measuring and managing, in Peter Drucker's sense of course, can't mean an entrepreneurial state using its funding to micromanage research projects. Rather, governments have to think in macroeconomic terms. To do this, the entrepreneurial state has to tip the scales with its economic weight much more often through innovation-friendly procurement.

139

THE ENTREPRENEURIAL STATE –
HOW CAN FRESH POLICY THINKING FOSTER INNOVATION?

 # 3.

Leverage the Purchasing Power of the State

DARPA has a competitive advantage that no other research funding institution in the world can even come close to offering: the purchasing power of the US military. The transition from prototype to series production, a process perfected over several decades, looks something like this, roughly speaking. When a DARPA-funded start-up has developed a new laser or stealth material to the point of serial production, the US Army, Navy, or Air Force orders a considerable amount. The order is placed at a time when the customer can't be 100 percent sure that the new high-tech company can actually deliver the product at the agreed time and in the defined quality. If there are problems, it's not an apocalyptic catastrophe for the supplier—the sales contract stipulates that this is exactly what might happen. The military procurement officials share in the risk, as DARPA's experience over the past few decades has shown. First, the development process is significantly accelerated once the customer has placed the order. Second, default rates are nowhere near as high as skeptical business executives might assume. The first volume order not only creates planning security for the supplier, it also increases the development teams' motivation and often makes it possible to develop the technology to the necessary maturity with an initial military lead customer and to reduce the price so that it's also interesting for a broader clientele in the civilian sector. Private investors love this two-for-one deal: financing of precommercial, risky product development by DARPA and a full order book are grist for the Excel mills of investment managers. And so the technology gains exactly the momentum it needs to transition to commercial exploitation.

Western Europeans (except for the British and French) and Japanese citizens are often wary of thinking about military and civilian innovation in parallel. There are many obvious reasons for this, especially in Japan and in the authors' home country of Germany. Following Russia's attack on Ukraine in February 2022 and in light of China's increasingly aggressive Taiwan policy, however, a process of rethinking has taken place. In recent decades, innovation has been a more important geopolitical factor than many pacifists (including one of the two authors of this book) have wanted to admit. Whether liberal democracies can compete against autocratic political systems or (in the case of Russia) against protofascist dictatorships from a position of strength in the coming years and decades will depend very much on their capacity for military innovation. This insight has also begun to take hold among the liberal-minded employees of major tech companies. In 2018, Google's AI researchers successfully rebelled against renewing a Pentagon contract for the Maven AI project that was contributing to the development of autonomous weapon systems.[26] Today, the company and its new Google Public Sector division are again competing for defense contracts—and employees no longer rebel. Viewed from a global perspective, this paradigm shift is overdue. Democracies that choose to defend themselves can learn a lot in this regard from Israel. For decades, close cooperation between public research laboratories, major governmental organizations, private defense companies, and the military has led to the development of highly innovative weapons technologies at record speed, ensuring Israel's continuing security. A particularly vivid example of this is Iron Dome, a mobile ground-based system to defend against short-range missiles, artillery, and mortar shells, which have been increasingly used in attacks on Israel since the mid-2000s. After the contract was awarded in 2007, the system was ready for use within four years. From the start, it successfully intercepted around 90 percent of enemy

141

THE ENTREPRENEURIAL STATE –
HOW CAN FRESH POLICY THINKING FOSTER INNOVATION?

missiles.[27] Since then it's been continuously improved, including with significant financial support from the United States.

For years, China and its surveillance technologies have provided an (alarming) blueprint of how an autocratic state can harness its purchasing power to promote disruptive technological innovation. The Chinese state places bulk orders for surveillance cameras, AI-based analysis software, new methods for processing biometric data, and facial recognition applications that can even identify people wearing masks. In this way, the government creates a huge domestic market that will spawn a number of national champions. In the next step, start-ups such as Hikvision, SenseTime, or Hanwang turn their highly innovative (and government-promoted) products into best-selling exports, preferably to other autocratic countries. Liberal democracies should draw the right conclusions from this.

Supporting innovation through defense tech will become radically more important in the coming years. In democracies, governmental innovation agencies' missions will include identifying as many dual-use cases as possible so that the innovative added value doesn't only go to military applications. This is the case at DARPA, where many program managers focus primarily on civilian applications. Sixty years of DARPA history also show how helpful it is for innovative leaps when state institutions not only see radical innovation as something that requires funding but also invest directly in highly innovative products as purchasers. In chapter 5, we'll describe in detail how the state and the market work together to create a funding cycle that helps innovators survive the so-called valley of death. This somewhat violent metaphor describes the phase in which it becomes apparent that an idea can turn into an innovative breakthrough, but investors hesitate because there are no customers in sight, and the path to a marketable product is always longer than hoped. For sensible practices of innovation-oriented procurement, we don't necessarily have to look to the United States or (in all its ambivalence) to China, where the state, as a purchaser of advanced technology aligned with its high-tech strategies, aids in the birth of entire industries. Midsized countries catch up despite smaller buying power because in market terms, government budgets are big.[28]

The French state is ordering cloud services that are still in the development phase from French providers on a grand scale, both from digital start-ups, such as New Vector, and from industrial consortia, such as Dassault Systèmes and OVHcloud with its government cloud solutions.[29] The Finnish government awarded the German-Finnish quantum start-up IQM a contract worth 20 million euros for a product that didn't yet exist: a quantum computer. UK Ministry of Defense also bought a quantum machine far from being market-ready from British start-up Orca.[30] With the German AI translation service DeepL, it was the Swiss federal administration that placed a large order early on and was happy to be named as a reference customer, which DeepL used in marketing.[31]

Interestingly, the EU and several European countries have already developed a procurement process for this, known as "precommercial procurement." Although it has existed for more than a decade, it's unfortunately not frequently used. Politicians interested in high tech have definitely gotten the message. To us, this is nice, but procurement oriented around breakthrough innovation should be envisioned on a much grander scale and beyond advanced digital technology.

Public spending is certainly well invested in education and health and infrastructure. But why don't governments of countries with high real estate prices and shortages of housing for middle-income families try to commission, say, half a million residential housing units at a price of 1,000 US dollars per square meter with appropriate environmental and fire protection standards and then rent them out in a socially responsible way? Perhaps we need to put this challenge

143

THE ENTREPRENEURIAL STATE –
HOW CAN FRESH POLICY THINKING FOSTER INNOVATION?

to innovators in the residential construction space rather than real estate speculators or municipal housing associations. An innovative building firm could obtain the properties inexpensively from communities, states, or the federal government. Or why don't governments in countries with slow adoption of e-mobility buy 100,000 charging stations for electric cars? Or 3,000 miles of highway with a cold- and crack-proof surface that doesn't use microplastics and doesn't have to be replaced every few years, as if planned obsolescence were written into the contract? Perhaps the public sector could commission a few skyscrapers or bridges that don't rely on reinforced concrete, or a couple of giant carbon capture facilities.

Let's remember the summer of 2020. Great Britain, the United States, and Israel quickly and unbureaucratically awarded AstraZeneca, BioNTech/Pfizer, and Moderna vaccine orders worth billions of dollars. Instead of building safeguards for themselves into the contracts, they largely released the manufacturers from liability for potential side effects. The British, American, and Israeli procurement officials didn't know at the time how effective and safe the vaccines were or whether they would ever be approved, but they rolled out a red carpet to innovative providers. In contrast, the European procurement bureaucracy approached the procurement process with its usual ultracautious mentality. As EU citizens soon realized, Europe paid dearly as a result of the EU Commission's innovation-hostile and short-sighted risk minimization strategy.

FINANCIAL RETURN

In a 2018 *Dilbert* comic titled "Winning the NASA contract," cartoonist Scott Adams slyly poked fun at the outsourcing of space technology to private companies. In this case, he targeted the lack of bidders on a contract that calls for contacting aliens (and that could prove to be a suicide mission).

As so often in his mini-stories of nerd life, Adams gets at the heart of one of the fundamental questions at the interface between government, the corporate world, technology, and society. In space travel, the principle of precommercial procurement has existed for decades and has become even more important in recent years as a consequence of the state space agencies' major outsourcing initiatives, collectively known as the "new space industry." The development of the Orion spacecraft and its crew module by Lockheed Martin and Airbus Defense—with all its ups and downs and delays—is a recent example that makes clear that neither the state nor the market working alone can bring about this kind of radical innovation. What would SpaceX be without NASA contracts? It requires an entrepreneurial state that understands market mechanisms and empowers market players to bring something radically new into the world. When designing the process for awarding contracts, however, governments have to ensure that competitive principles will apply. If the purchasing power of the state turns into a de facto subsidy for a few big players—for example by issuing extremely complicated invitations for bidding—that reduces competition and artificially raises prices, then the effort will backfire. Innovation will suffer at taxpayer expense. Unfortunately, this has happened over and over again. In Europe, for example, the subsidization of rooftop solar roofs in the 2000s led to prices remaining artificially high and to European manufacturers' loss of competitiveness, which ultimately let China roll up the market with cheap rooftop solar panels.[32] One good thing did happen: the prices just kept falling. It seems that Europe is now making the same mistake again, in particular with heat pumps. Instead of priming the market with precommercial procurement of the most affordable heat

145

THE ENTREPRENEURIAL STATE –
HOW CAN FRESH POLICY THINKING FOSTER INNOVATION?

pumps, the usual players are enjoying major subsidies that constitute a large percentage of total costs.

If countries or communities of nations stimulate the development of new technologies at the right time while preserving market mechanisms and with unbureaucratic award processes, they can help these technologies survive the valley of death of innovation.

It's striking how the state can play this role as bridge builder over innovation's valley of death better than the market by itself for three reasons. First, the state enjoys a much lower cost of capital than private investors. The riskier the investment in innovation, the greater the difference is. In the medium term of up to ten years, governments around the world can realistically invest capital at an interest rate of 2 to 3 percent. Venture capitalists demand at least 10 percent and additional "leverage," if they're willing to invest their capital over an adequately long term at all. This brings us to the second point. Real innovative leaps usually require a development time measured not in years but in decades. Only the state has that kind of perseverance, and only it can help companies develop it. But third, and by far the most important difference between the state and the capital market when it comes to investing in innovative breakthroughs, is the financial return model. The capital market only measures returns by one thing: profit, the increase of money invested, as quickly as possible. If the entrepreneurial state invests in innovation in a targeted and mission-oriented way, it can be good business in purely monetary terms.

And of course, innovative companies pay taxes that can be redistributed or reinvested by the government. But the returns to the state and society are much more varied than those of venture capitalists. But also the profits inside those companies are often redeployed for research and development of new products, as again the Moderna and BioNTech examples show, as they are now developing individualized cancer treatments. From the perspective of society, they also improve education and create good jobs. They can help us to achieve environmental and climate goals and to bring about an overall higher level of prosperity. And ultimately, innovation can increase social happiness and lead to greater social participation by all groups in society.

This is only an approximate equation, unfortunately. As far as we know, there's still no comprehensive model of socioeconomic returns that would enable a quantified overall view of the "return on investment" from government-funded and government-induced innovations. Creating such a model strikes us as a difficult but worthwhile undertaking for innovation research in the next few years. For objectives and methodology, researchers could draw inspiration from attempts to quantify the socioeconomic returns of investment in education. In the case of education, we have methodologically sound empirical proof that investing in early childhood education has a big economic payoff.[33] With a model like that for innovation, politics and society would have a better view of and more compelling justification for why it's worth investing in innovation.

In investment theory, innovation is an option. For a better future that we can work toward today, we need more options. Something will always go wrong along the way. Sometimes it's even funny. After Running Man, the rescue robot from Florida, had brilliantly completed all the obstacles on the course and performed its victory dance, the machine unfortunately lost its balance and landed on its face with a metallic clang. No points were deducted for the fall.

147

THE ENTREPRENEURIAL STATE –
HOW CAN FRESH POLICY THINKING FOSTER INNOVATION?

Rescue robots don't have to be able to dance. Some of Running Man's specialized peers are much better at it.

Take a look at the Boston Dynamics channel on YouTube under the heading "Do you love me?"

FINANCING BREAK-THROUGHS

HOW CAN NEW TECHNOLOGIES SURVIVE THE VALLEY OF DEATH?

THE FIRST TECH INVESTOR

The actual plan of Johannes Gensfleisch was to get rich selling pilgrim mirrors. The way his products worked is that, by using mirrors made of an incrementally innovative tin-lead alloy, people could collect the benevolent aura emitted by saints' relics and bring it home with them from a pilgrimage. Dissatisfied customers would likely have had a hard time proving the product was defective. But even with such a sure thing as this, Gensfleisch's plan for getting rich didn't work out. An epidemic postponed the scheduled pilgrimage, and the market collapsed. The inventor, developer, and entrepreneur made a new attempt a few years later, and once again, the faithful were the target group. From a purely financial point of view, the second major venture of Johannes Gensfleisch (more widely known today as Gutenberg) wasn't a great success either. He had a falling-out with his investors, and that's rarely a good idea for start-up founders. But his 180 Bibles, printed between 1452 and 1454 with movable metal type produced with an innovative hand-casting instrument using new alloys, an improved printing press, and an improved formula for oil-based printing inks, made Gutenberg one of the greatest breakthrough innovators in world history.[1]

The birth of printing is still instructive in many ways because it shows how disruptive technologies come into the world under trying circumstances. The US publicist and internet pioneer Jeff Jarvis calls Johannes Gutenberg the first tech entrepreneur and the "patron saint of Silicon Valley." While that may sound overly dramatic, this formulation is essentially correct, although primarily in relation to an aspect that Jarvis pays relatively little attention to in his essay *Gutenberg the Geek*, that is, securing financing in anticipation of markets and business models that don't yet exist.[2] Gutenberg's most important financial backer was the wealthy and belligerent businessman Johannes

151

FINANCING INNOVATION LEAPS –
HOW CAN NEW TECHNOLOGIES SURVIVE THE VALLEY OF DEATH?

Fust. To follow Jarvis's logic, tech investor Fust would be the patron saint of Sandhill Road, the street in Silicon Valley where the major venture capital firms have their offices today. But first things first.

The historical sources for the life of Johannes Gutenberg are unfortunately incomplete. We know that he was born around 1400 as the third child of a patrician family in Mainz. It's highly likely that he spent part of his youth in Eltville. He may have studied in Erfurt. Where did he live as a young man, what did he do in his twenties, what technical skills did he acquire and how? The historians have no answers. But it is certain that from 1434 to 1444, Johannes Gutenberg cut and polished precious stones in Strasbourg and earned money as a coin minter and goldsmith. During this time, he also must have learned to work with metal and glass, because in the mid-1430s he got involved in his first known *aventur*. At that time, an *aventur* was the name for a high-risk venture that was cofinanced with outside capital. The capitalists—a term that didn't exist at the time, of course—were usually wealthy townspeople, although members of the nobility were also involved as venture capital investors. The Latin root of *aventur* is *advenire*, which means "to come, to reach, to arrive." The path between departure and arrival is the adventure, and it's what venture capitalists do when they place their bets on tech start-ups.

Gutenberg's first well-known business venture in Strasbourg was the production and delivery of 32,000 pilgrim mirrors made from a novel lead-tin alloy. Hans Riffe, feudal administrator of the town of Lichtenau, provided the venture capital, although the risk likely seemed minor; the technically experienced Gutenberg had evidently mastered the production technique better than anyone else. The large-scale production of mirrors was intended for the great Aachen pilgrimage of 1439. To those involved, the project seemed like a sure thing. The *aventur* didn't run into trouble because of poor product quality. And the credulous believers probably would have bought the

GUTENBERG WAS CONVINCED THAT HIS BIBLES WOULD FIND CUSTOMERS BECAUSE AS PRODUCTS OF THE PRINTING PRESS, THEY WERE RADICALLY LESS EXPENSIVE THAN MANUSCRIPTS.

well-made mirrors. Unfortunately, however, the pilgrimage had to be postponed for a year because of the bubonic plague, resulting in a legal dispute between Gutenberg, Riffe, and the heirs of a third business partner, Andreas Dritzehn, over who was owed repayment of the capital for the aventur, and how much. The pattern of strained investor relations would be repeated with the serial founder and innovator Gutenberg—and again with an innovative leap that wasn't based on ecclesiastically tolerated superstition, a breakthrough that would pave the way for the information revolution of the Enlightenment.

In October 1448, Johannes Gutenberg returned to his native Mainz and borrowed 150 guilders from a cousin, which he invested in setting up a printing workshop. In the middle of the fifteenth century, 150 guilders was a lot of money. A fattened ox cost 4 guilders. But for Gutenberg, that was more of a seed round, as it would be referred to in the start-up scene today: a small, early investment that plants the seed for something big. By 1449, Gutenberg had clearly made significant technical progress with his experiments in book and broadside printing. He convinced the merchant Johannes Fust to invest 800 guilders, about what a large house for a wealthy patrician in Mainz would cost, in this new *aventur*. This represented a much riskier adventure than a loan to manufacture mirrors for collecting the aura from saints' relics in times of bubonic plague, because at that time, there were hardly any buyers for printed texts. Or, to put it in modern terms, the market that is typically present for innovative breakthroughs didn't yet exist.

By far the largest customer for books was the Church. It employed scribes of its own and was neither in a hurry to copy books nor was

153

FINANCING INNOVATION LEAPS –
HOW CAN NEW TECHNOLOGIES SURVIVE THE VALLEY OF DEATH?

it subject to austerity constraints. Books, usually handwritten Bibles, were made to order. The state—that is, monarchs, electors, and royal families—occasionally ordered records of various kinds, which were mostly unique in each case and wouldn't have been worth printing. Universities, perhaps the most interesting target group and potentially open to innovation, were still small and few in number, and they had limited budgets. The risk undertaken by the Mainz patrician Johannes Fust really did take vision and courage. In second-round financing in 1452—on Sandhill Road, you would call it Series B—he doubled his stake in Gutenberg and anted up another 800 guilders. With this second financing round, presumably negotiated as a combination of credit and investment, the entrepreneurs and investors hoped to produce the *Werck der Bücher* (the "work of books"): the 180 famous Bibles with forty-two lines per page. Over the centuries, scholars in history and media studies have filled shelf after shelf with secondary literature on the printing and design of the B-42 edition. But little thought was given to the innovative business model. It too was disruptive because, unlike manuscripts, the Gutenberg Bibles didn't enjoy the financial safeguard of an advance order by a paying customer. Who buys a product that they haven't even seen?

Gutenberg and Fust were convinced that their Bibles would find customers because, as products of the printing press, they were less expensive than manuscripts. Incidentally, the typography and page design of the Bibles were closely based on manuscripts so that the new product didn't seem *too* disruptive. But that didn't change anything about the fact that there was no proof of concept for either the product or the business model, that production would take at least two years, and that no one had any way of knowing whether the technology would live up to its promise in series production. As we know today, the technology more than lived up to its promise, the innovator and his team of around twenty employees did work of consistently good quality, the Gutenberg Bibles found their customers, and thanks to Johannes Fust's sufficient capital reserves, the innovative leap of the printing press survived the phase of development that innovation researchers often refer to today as the "valley of death."

TECHNOLOGY READINESS IN THE VALLEY OF DEATH

How do innovators finance the long journey from an idea to the successful marketing of a new technology in the form of products and applications? The history of innovation has spawned numerous models since the printing press.

With their model combining frugality and self-exploitation, known as "bootstrapping" in start-up jargon, Gutenberg and Fust were way ahead of the "venture capital investment" of their time. In the Italian Renaissance, the patronage model for innovative minds was predominant. The Medici gave Galileo Galilei the financial freedom to rethink the world. The initial situation of many gentleman scientists of the eighteenth and nineteenth centuries, such as the polymath and American statesman Benjamin Franklin, the naturalist Charles Darwin, or the discoverer of hydrogen, Henry Cavendish, was just as comfortable.[3] They had been born into wealthy families or achieved economic independence as businessmen and didn't need to look for patrons. Some of the great inventors in history, such as the forestry official Karl von Drais, faced negligible demands on their time from their principal occupation so they were able to devote considerable time to their true calling. Drais was a serial inventor whose innovations include a formula for approximating the solutions to numerical equations of any degree, a typewriter, a stove that burned wood sparingly, and the "running machine," the oldest ancestor of the bicycle, which has made a brilliant comeback over the past twenty years in the form of the balance bike for young children.[4]

155

FINANCING INNOVATION LEAPS –
HOW CAN NEW TECHNOLOGIES SURVIVE THE VALLEY OF DEATH?

The rise of universities and funding foundations, military research and government laboratories, research and development departments in large corporations, and also the financial innovation of the venture capital fund have institutionalized the funding and process of creating new things, especially in the vast field of digital innovations.

In addition, legal innovations over the centuries now give technological innovations some financial breathing space. These include joint stock companies and corporate structures with limited liability as well as intellectual property rights. Legislators indirectly fund innovation by exempting drug manufacturers from liability for side effects (as the UK did recently with the AstraZeneca vaccine). Or they promote a new technology through apportionment procedures at the expense of consumers, such as Germany's Renewable Energy Act.[5]

And yet, despite the variety of models and approaches, the fundamental chicken-and-egg problem faced by breakthrough innovators has remained the same. Who will buy a technology, an application, or a product that doesn't yet exist? What supplier, in the absence of orders from customers or retailers, will take the risk of bringing the technology, the application, or the product to a market that doesn't yet exist?

With a closer look at innovation's "valley of death," it is clear that innovative procurement policies can help, especially in the middle phases of the technology development process, because in these phases the uncertainties affecting decision-making are greatest for pure market players along at least four dimensions:

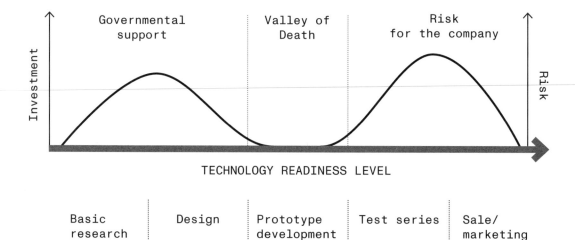

Investment

Governmental
support

Valley of
Death

Risk
for the company

Risk

TECHNOLOGY READINESS LEVEL

| Basic research | Design | Prototype development | Test series | Sale/ marketing |

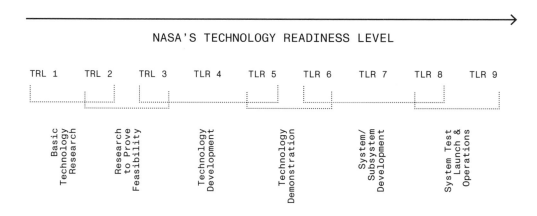

NASA'S TECHNOLOGY READINESS LEVEL

TRL 1   TRL 2   TRL 3   TLR 4   TLR 5   TLR 6   TLR 7   TLR 8   TLR 9

Basic
Technology
Research

Research
to Prove
Feasibility

Technology
Development

Technology
Demonstration

System/
Subsystem
Development

System Test
Launch &
Operations

157

FINANCING INNOVATION LEAPS –
HOW CAN NEW TECHNOLOGIES SURVIVE THE VALLEY OF DEATH?

1. Technical feasibility
2. Speed of the development process prior to market readiness
3. Applicability to an actual problem (specificity of technology)
4. Price at which the new technology can be offered profitably

If the state makes advance payments as a purchaser, the risks decrease in dimensions two through four. With a secure budget, development is delayed less often and less severely. There are already use cases for which the product can be specifically developed. And with the first large order, the first economies of scale take effect and reduce unit costs. What remains, of course, is the technological risk. Maybe in the end the technology won't work after all. This risk can't be avoided, because otherwise it wouldn't be a potential technological innovation. But what does risk mean in this context? Every risk needs to be considered from the contrasting perspective of the status quo—and the risk of not improving it.

National governments spend hundreds of billions of dollars to promote new technologies in basic research and under the label of "application-oriented research" up to the point where entrepreneurial researchers or researching entrepreneurs enter innovation's valley of death. That's where the support all but comes to an end. To us, this seems like an economically absurd reversal of the "sunk cost bias" mentioned earlier: We resist admitting the truth to ourselves when we've made a bad investment. Instead of drawing a line through the balance sheet and declaring the mourning period over for the money we threw away, we often want to inject even more capital, even though facts are telling us to do otherwise. As governments, societies, and economic systems, we follow the reverse logic when it comes to promoting innovation. We choke off funding at the very moment when it would actually make sense to continue investing, because the chance of success increases as development matures.

With eyes wide open, we're throwing away piles of money because we don't want to take any more risks, even though the biggest risks are already behind us. Or to put it figuratively, we're letting the bird in hand starve.

In conversations with breakthrough innovators, this metaphor takes concrete form much more often than an innovative society would want. Up through completion of a doctoral thesis, funding is no problem at all for the talented and hardworking, even if it comes with highly inflexible conditions. It gets exceptionally difficult just when hard-won knowledge is to be turned into valuable applications. Until the "proof of concept" point is reached, researchers with entrepreneurial ambitions often find the energy to somehow find financing. But at technology-readiness level three or four, their reserves are exhausted. Bootstrapping can't get them any further. The dry stretch leading to levels seven, eight, and nine is long, and too often it is too long. In essence, this is a case of market failure. More precisely, capital market failure, and even more precisely, venture capital market failure for breakthrough innovations and deep technologies. To solve this problem, history provides a blueprint: the venture capital system as it used to be.

GETTING THE FUNDING CYCLE ROLLING FOR DEEP TECH

In 1946, just after World War II, physicist and MIT president Karl Compton, French American Harvard economist and military planner Georges Doriot, and two additional partners formed a private equity firm with the unremarkable name of American Research and Development Corporation. In contrast to its name, the goal of ARDC was quite specific: to provide veterans returning from the war readier access to capital so they could take the path of self-employment and

159

FINANCING INNOVATION LEAPS –
HOW CAN NEW TECHNOLOGIES SURVIVE THE VALLEY OF DEATH?

found their own companies. ARDC achieved solid growth for a few years without attracting much attention. With the passage of time after the war, the goal of funding veterans naturally receded into the background. Compton and Doriot ultimately went down in the history of finance for a different reason. In 1957, they invested $70,000 in a new Boston computer company named Digital Equipment Corporation, securing for themselves an outrageous (from today's perspective) 70 percent stake in the company. DEC went public in 1968 with a value of $35.5 million. The investors had multiplied their initial stake 350-fold.[6] ARDC was thus the first major venture capital success story of the digital age, sending two messages to investors around the world with an appetite for risk. First, the future belonged to computers. And second, there were gigantic bets to be won.

ARDC's big score with DEC played a key role in ensuring that smart and competent founders in the digital innovation space enjoy relatively easy access to capital today, sometimes even to excess, at all stages of start-up development. Initially in the United States, later in Asia and Europe, and recently in Africa and South America as well, venture capital ecosystems have emerged that have set a flywheel of corporate financing in motion. In an ideal case, one turn of the flywheel looks like this.

Business angels and start-up incubators make it possible for founders to develop their digital ideas, build prototypes, and test them. A lot of ideas turn out to be nonsense. But if the test results are encouraging, early-stage investors get involved by funding product development and market positioning up to a point where the chances of success for scaling the product or service are significantly increased. At this point, the growth financing providers take over. Depending on the industry or business model, that might mean Series B or C. As a rule, investors in this middle phase implement a strict regime of performance metrics to measure growth paths against the

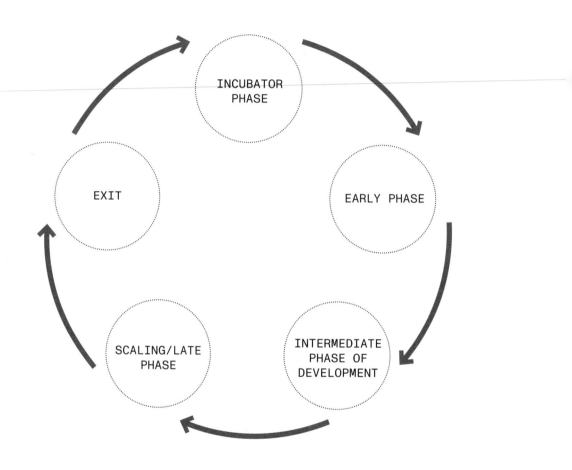

161

FINANCING INNOVATION LEAPS –
HOW CAN NEW TECHNOLOGIES SURVIVE THE VALLEY OF DEATH?

capital employed. If the start-up at least approximately achieves the growth that it promised in its optimistic business plans illustrated with exponential curves and the costs don't get out of hand, then late-stage investors will ultimately provide capital to prepare the exit, ideally via an initial public offering like DEC's.

The risk–reward calculation along the financing chain is in principle very simple. The earlier the investment, the higher the chances that returns will be many times greater than the initial capital—but also the higher the risk of a total loss. The later investors get involved, the more precisely they can calculate their risk–reward ratio based on the company's performance and market data. The rise of the digital venture capital scene—and its successes as more and more billion-dollar gambles have paid off—is closely linked to investors specializing in each individual phase. Angels and incubators have developed an eye for major talent and the right ideas. Late-stage investors can calculate down to the last decimal point how much impact their capital might have, under which conditions, at which burn rate, for which level of market success. The decisive factor for the flywheel logic of this system is having competent people making decisions at every point and ensuring that there's sufficient capital for promising founders in every typical development phase. When that succeeds, as it has in Silicon Valley since the late 1960s, a self-reinforcing system emerges.[7]

Every exit releases capital that not only makes the founders rich and lets investors (or at least the self-promoters among them) buy new yachts. A large part flows back into the system, allocating more capital for start-ups that hope to create new digital innovations. The next generation of digital founders will then have better chances of successfully crossing innovation's valley of death, or at least not failing due to a lack of capital. Over the last five to ten years, the excesses of the digital innovation venture capital system, with its exaggerated valuations and short-term profit expectations lacking all basis in reality, with its herd mentality and perpetual focus on chasing the same copycat products and platform-based business models, have become obvious. The logical consequence was the stark market correction of 2021 and 2022. It's understandable if sympathy is limited for those whose market expertise didn't match their outsized expectations.

But looking at it in terms of macroeconomics and financial history, it's also clear that the invention of venture capital by Compton, Doriot, and others was an innovation in the capital market at just the right time that made the development of digital technologies possible. Traditional financiers, such as banks and private investors, weren't willing to take the necessary risks because they didn't see the potential opportunities. We're currently experiencing a similar situation for the financing of breakthrough and deep-tech innovations. Here again, financial market innovation is needed. In a very literal sense, however, it's not necessary to reinvent the wheel. With minor adjustments, the flywheel model for light asset innovations can be adopted for the type of innovation that the world needs more urgently than the next quick commerce app or feedback tool for the human resources department. What it does require is a different mindset in the traditional venture capital scene when classifying opportunities and risks.

"Market risk is inversely proportional to technical risk," as the saying goes. The statement is attributed to venture capital investor Tom Perkins and is sometimes known as Perkins's law.[8] We're not sure that this observation should actually be treated like a law of economics because there are too many exceptions. But at the very least, the saying well describes how the venture capital scene has sized up technologically driven innovation, especially in the last ten years. If a start-up wants to achieve quick growth based on a largely mature, modestly innovative technology in connection with an innovative business model, investors face a high risk of failure. It's a natural

163

FINANCING INNOVATION LEAPS –
HOW CAN NEW TECHNOLOGIES SURVIVE THE VALLEY OF DEATH?

consequence of the competitive situation. What one start-up can do, others can too. A ridesharing app like Uber can try to conquer the traditional taxi market. But the technology and the business model are easy to copy. As a result, Uber, Lyft, Didi, Bolt, MyTaxi, and many others can end up competing with each other so fiercely for years on end that for the foreseeable future, none of them will make money, and most of the active market players will still fail even after making extremely high initial investments. Late-stage investors in particular are at risk of burning through billions of dollars.

The prototypical case of the reverse situation with high "technological risks" can be observed in the pharmaceuticals market. The risk that a potential new blockbuster drug will fail to gain regulatory approval is extremely high. But if the drug is safe, its active ingredient actually works as desired, and it clears all regulatory hurdles, market success is very likely, and direct competition is significantly reduced by patent protections, among other things. The defining understanding of risk among digital venture capitalists over the past decade has been a clear preference for betting on markets instead of betting on technology. The basic idea might be expressed more or less as follows: if a start-up isn't growing, we'll just throw more money at the problem. Until mid-2021, betting on the market paid off often enough. The bursting of the tech bubble in early 2022 dampened the mood in the casino, though.[9] The good news for breakthrough innovators is that venture capitalists increasingly understand that betting on technology is much more attractive than they previously thought. Truly innovative technologies may need a bit more time on average to climb the ladder of technology readiness. But superior technology creates lasting competitive advantages—and it comes with the possibility of conquering and holding markets without a need for constant marketing battles or ruinous price wars. And deep tech founders in many areas don't all bear a similarly high technology

MARKET RISK VS. TECHNICAL RISK

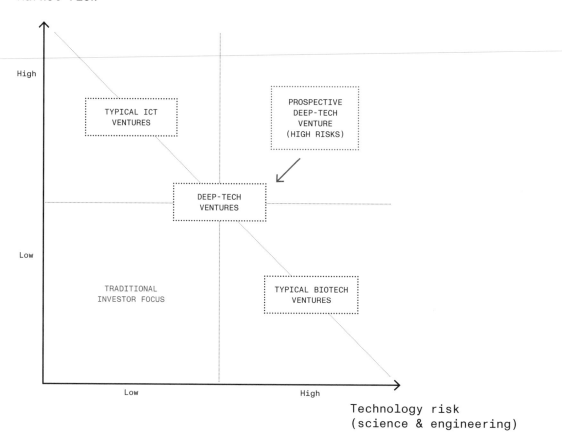

Source: BCG and Hello Tomorrow

165

FINANCING INNOVATION LEAPS —
HOW CAN NEW TECHNOLOGIES SURVIVE THE VALLEY OF DEATH?

risk as in traditional pharmaceutical development, where unexpected study results during a highly complex approval process can stop a new active ingredient in its tracks. With deep tech, it's often possible to gauge which methods will be needed to overcome likely hurdles. And for innovative materials or new robots, the regulatory risks are of course lower than for new drugs.

BETTER BETS FOR ALL

It's high time for venture capital financing to shift from the easy "digital" asset classes to deep technologies. And it's also a good time to get the funding cycle rolling for deep-tech companies, whose technologies and business models almost all contribute to at least one of the UN Sustainable Development Goals.

The bets on so-called asset-light companies, such as software-as-a-service providers or specialized online retailers or platforms, are significantly riskier than they were in the 2010s because of increasing competition and low levels of innovation. Making a sufficient number of bets and then topping them off with enough money no longer works when a company's valuations get so high that there's no way its earnings will ever justify them, even if the business continues to grow at a healthy rate. At the same time, there's a lot of excess money sitting around in the venture capital system that could and should be invested, but investors are waiting for interesting investment opportunities. In industry jargon, this money is known as dry powder. The management consultancy BCG and the French deep-tech think tank Hello Tomorrow estimated the dry powder risk capital held by venture capital and private equity companies in spring 2021 at $1.9 trillion.[10] Since then, it's highly likely that this sum has increased significantly. Despite rising interest rates, it faces extreme inflation risk. Dry powder can be invested a lot more intelligently and with far greater impact on progress than the next foie gras start-up that will supposedly make our lives a little easier. The good news is that rethinking—and shifting investment positions—has already begun.

According to the BGC/Hello Tomorrow study, venture capital investments in deep-tech start-ups quadrupled to more than $60 billion between 2015 and 2020. A significant part of this is investment by the major corporates, which invested almost $20 billion in deep-tech start-ups in 2020. Of the total, the lion's share, around 80 percent, went to the areas of synthetic biology, AI, and advanced materials.

The authors of the study estimate (and hope) that the deep-tech growth trends for venture capital will continue unabated in coming years, with more than $200 billion flowing into this asset class by 2025. Optimistic estimates of this type in studies by consultants are of course always highly uncertain, even if slick infographics seem to suggest these developments are inevitable. Naturally, we don't want to equate deep-tech investment with breakthrough innovations for progress, as we defined them in chapter 2. But in principle we consider this asset class and the recognizable trends in this market to be a good proxy for how sufficient capital can be made available to science-based start-ups with happiness-maximizing development potential so that the founders' financing risks can be minimized in innovation's valley of death.

In view of the gigantic sums of risk-embracing dry powder waiting to be invested, the forecast of $200 billion in deep-tech capital strikes us as a plausible extrapolation, taking current market trends into account. But compared to the trillions that flow into digital and asset-light start-ups worldwide each year, it's only a trickle. If all the stakeholders involved in risk capital financing thought bigger and started making the right changes in the right places in the capital system, a lot more might be possible with a financing flywheel for disruptive innovation.

167

FINANCING INNOVATION LEAPS –
HOW CAN NEW TECHNOLOGIES SURVIVE THE VALLEY OF DEATH?

DEEP-TECH INVESTMENT

DEEP-TECH INVESTMENT IS UNEQUALLY SPREAD WITH AROUND 80% ACCOUNTING FOR SYNTHETIC BIOLOGY, ARTIFICIAL INTELLIGENCE, AND ADVANCED MATERIALS

Deep tech total investments by
technology in 2020 ($B)

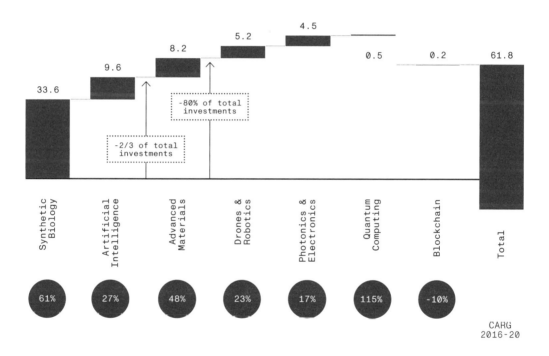

Note: investments include private investments, minority stakes,
initial public offerings, and M&A; transactions mapped on several
technlogies were split equally between these technologies;
25–30% of transactions remain undisclosed

Source: Capital IQ, Crunchbase, Quid, BCG Center for Growth & Innovation Analytics,
BCG and Hello Tomorrow analysis

Increase the Deep-Tech Competence of Venture Capitalists and Private Equity Firms

After pitching their ideas to venture capitalists many times, deep-tech breakthrough innovators have told us over and over again, "The interest was fundamentally there. But the financing ultimately failed because the decision-makers lacked the expertise to assess the technology, and they admitted it." Anyone placing technology bets inherently needs to understand the technology and the development risks associated with it, especially in the early stages of the investment cycle when market data aren't yet available. So far, the teams working at venture capital funds and private equity firms have as a rule been weighted too heavily toward generalists to be able to competently take advantage of the tremendous opportunities offered by truly disruptive technologies. To change this, they need to (a) bring scientists with an interest in financing issues directly into their teams of analysts and open up the partner track for them, and (b) create networks with external consultants who aren't drawn from the usual suspects of the tech consulting scene but are instead active researchers in the field in which the fund hopes to invest. What makes things more difficult for private equity companies is that deep tech is too often equated with early stage and thus doesn't fit into the investment patterns of most private equity firms. This view is already outdated. Numerous deep-tech start-ups are in the second half of the funding cycle right now. In the years to come, tech-savvy private equity firms will take a closer look at this area. Meanwhile, the non-tech-savvy firms are going to miss out on major opportunities for their investors.

RISK INTELLIGENCE ALSO MEANS UNDERSTANDING WHEN NOT TAKING ANY RISKS IS A MAJOR RISK.

169

FINANCING INNOVATION LEAPS –
HOW CAN NEW TECHNOLOGIES SURVIVE THE VALLEY OF DEATH?

Increase the Long-Term Thinking and Risk Intelligence of Corporates and Family Offices

Venture capitalists have gotten used to quick wins—or to knowing relatively quickly if they've made a bad investment. If they're successful, it usually takes no more than seven years from seed financing to exit for software-as-a-service applications, e-commerce, or platform business models in business-to-consumer markets. With deep-tech start-ups, investors have to wait more like ten to fifteen years and sometimes two decades before the companies actually reshape their markets. The big difference, however, is that digital innovators often only claim to be disruptive. Most of the time, they're actually just discovering new markets—and often just new market niches. They're much less of a threat to established companies than the term "disruptive" suggests. Breakthrough innovations are disruptive by definition. The bigger risk is not being invested in a potential breakthrough. This is especially true for the large tech corporations that were founded before the invention of the internet. Major chemical, automotive, engineering, medtech, pharmaceutical, and infrastructure corporations are rapidly increasing their mergers and acquisitions activity in deep-tech start-ups, which is an encouraging sign of increasing risk intelligence among these key players. When it's your own market being disrupted, you really need to be part of the action.

A similar development can be observed in many family offices. For them, long investment horizons are often less of a deterrent because they're accustomed to thinking in terms of generations. Another positive development that's been discernible for some time is that investment decision-makers are increasingly focusing on what is referred to as the potential life cycle impact of start-ups. This means that an investor's decision isn't primarily based on whether they can realize a profit after the next round of financing or take their cut on exit at the latest, but whether they're investing in a company that might remain in their portfolio for a very long time. Both corporate and family offices have great potential to radically improve the financing of disruptive innovations. But for that, the legal conditions have to be right.

The Right Government Conditions

Imagine a world where governments provided all the venture capital anyone could wish for. It probably wouldn't be very innovative. There would be no need to develop a risk capital market with all the competitive mechanisms that systemically promote innovation. Instead, there would be a subsidy-based economy whose rewards would predominantly go to people who knew how to simulate innovation using the cultural techniques of innovation theater in order to secure more subsidies for themselves. In chapter 4, we argued that the entrepreneurial state should be committed to playing an active role. But in the context of the (partial) market failure in financing breakthrough innovations at technology readiness level 3 or 4 and above, the government response can't be to flood the market with capital. That wouldn't be efficient or effective. Governments need to carefully incentivize the stages in the funding cycle where the flywheel loses momentum. For example, if an economy is experiencing a discernible shortage of growth capital for deep-tech companies at Series B and beyond (as is currently the case in Europe), then the government should either target investment at this stage with a sovereign wealth fund or create a quasi-public deep-tech fund of funds for growth capital that would allow the government to mobilize relatively large amounts of private capital with relatively little public capital. In the long term, this investment strategy will most likely pay off for taxpayers and government finances. Because where there's a lack of capital due to market failure

171

FINANCING INNOVATION LEAPS —
HOW CAN NEW TECHNOLOGIES SURVIVE THE VALLEY OF DEATH?

(for example, because of bias against betting on technology or too much focus on short-term returns from a generation of venture capital fund managers trained in the 2000s and 2010s), chances of especially good deals are higher.

Another effective governmental tool in this scenario consists of waterfall funds. In this type of fund, the state assumes part of the default risk. In exchange, the return is capped, and, if successful, the state collects the profits above the cap. This creates a major advantage in that this asset class is also attractive for institutional investors. Like pension funds, institutional investors are oriented toward investing on a very long term, and with their broad diversification, deep-tech investments would actually be very interesting for them, but they're not allowed to invest in this risk category because of an incorrect legal understanding of risk. Particularly innovative countries should have the confidence to set up something like a 401(k) plan based on an appropriate mix of investments in breakthrough innovations. This might mean an enormous leap forward for a country's innovators and at the same time make pensions much safer than they are today in the aging societies of Europe, Japan, South Korea, and China, where often systems based on taxes and public reapportionment are the rule. Sweden's equities-based pension system has already benefited significantly from the success of the Swedish start-up scene and could provide a promising model for how old-fashioned bureaucratic welfare states can transform themselves into entrepreneurial states who keep an eye out for start-up innovation and its economic potential.[11] It remains a mystery why investments in legacy industries are more favored by many countries' tax codes than investments in the technological future. And even if it's not primarily a governmental responsibility, political support for the establishment of a tech stock exchange would help in Europe. For European technology start-ups, it's always a long road to exit via initial public offering. They either have to successfully achieve a listing on the general stock exchanges, or choose the direct path to NASDAQ and the United States. A European tech exchange would give the flywheel for European technology start-ups a last, decisive push and release the funding necessary for all the previous phases. It would also keep capital and value creation in Europe. Today, from a European perspective, added value migrates far too often to the United States, and also increasingly to China and the Gulf States.[12] For these countries, a European tech exchange would actually represent an interesting area of investment.

Inclusive Venture Capital

In an essay entitled "Why Can't Tech Fix Its Gender Problem?" in the *MIT Technology Review,* the American historian Margaret O'Mara provides a striking and systematic overview of the many historical and cultural factors making the US tech industry not at all like the gender- and color-blind meritocracy it claims to be.[13] It describes the gender problems in STEM education, the tech geek stereotypes, and how hiring by employee referral with its self-reinforcing social selection tends to create monocultures. She shows how the underrepresentation of women has often been blamed on women themselves. And then O'Mara gets to her actual point: "What really lies at the core of tech's gender problem is money." Or more precisely, "Venture capital investment has been and remains the tech ecosystem's least diverse domain."

She also provides current figures. "White and Asian men make up 78% of those responsible for investing decisions and manage 93% of venture dollars overall." While there are, according to O'Mara, more female-led investment funds than there were a few years ago, "the majority

WHAT REALLY LIES AT THE CORE OF TECH'S GENDER PROBLEM IS MONEY.

173

FINANCING INNOVATION LEAPS –
HOW CAN NEW TECHNOLOGIES SURVIVE THE VALLEY OF DEATH?

of venture capital firms still have zero women as general partners or fund managers. Of the few women in these roles, nearly all are white. The US venture capital industry invested a record-breaking $329 billion across more than 17,000 deals in 2021. Only 2% of this bonanza went to start-ups founded solely by women—the lowest level since 2016. Less than 0.004% of the venture capital invested in the first half of 2021 went to start-ups with Black female founders." Extreme inequality in venture capital isn't only an American problem. A look at the European venture capitalist female founders dashboard again shows clearly lower rates of financing of female founders. Not even 1 percent of venture capital went to start-ups with all-female founders. As might be expected, the negative bias affects not only women but also people of other races and ethnicities. According to data from Crunchbase, Black founders in the United States only get around 1 percent of the venture capital pie.[14] And even if there has been welcome growth in venture capital for African founders since 2020, access to venture capital in the Global South remains at a level that corresponds to every postcolonial cliché.

Of course, we also don't have a simple cure for how such an obviously male-dominated decision-making process, in which the principle of pattern recognition gives such an obvious preference to men, can finally be made more inclusive. Indeed, according to Kleiner Perkins partner John Doerr, the most successful founders "all seem to be white, male nerds who've dropped out of Harvard or Stanford, and they absolutely have no social life."[15] The increasingly insistent discussions of this topic, scientific studies on the effect of similarity bias in investment decisions, and the success stories of female and non-white founders can only represent the start of gradually improving the situation. But we can hope that the paradigm shift toward deep tech and impact investing also offers an opportunity for critically questioning gender and racial bias. Breakthrough innovations need the input of diverse teams. That includes a diversity of founders, and it requires inclusive funding. Ultimately, the mostly still white and male decision-makers will have to come to the realization that if they continue to rely primarily on white male founders, they will limit their own investment opportunities. There will be better deals to be found with creative women and non-white founders.

The women in tech who today appear on the lists of the world's richest people are either the widows of tech tycoons, like Lauren Powell Jobs, or their ex-wives, like MacKenzie Scott, formerly married to Jeff Bezos. If one day the widowers and ex-husbands of female tech founders appear in these lists, then we'll know we're on the right track.

BETTER BETS FOR ALL

Joseph Schumpeter, who coined the term "creative destruction" and developed its conceptual model, emphasized the importance of investors for the innovation process and explored the implications of their role, even in his early works written before World War I. Ideally, according to Schumpeter, banks should offer disruptive entrepreneurs long-term loans with interest rates on market terms, thereby promoting the market economy's logic of progressive development. In the 1920s, the Austrian American Schumpeter, now a professor at Harvard, was able to observe this model in practice on both sides of the Atlantic. Cultures of innovation were flourishing wherever there were strong scientific institutions and sufficient capital for innovators from banks: in the United States and the UK, in Germany and France, in Scandinavia and in Switzerland. Around 1940, after enduring personal misfortune and the outbreak of World War II, Schumpeter took a dimmer view of capitalism as a force of renewal.

175

FINANCING INNOVATION LEAPS –
HOW CAN NEW TECHNOLOGIES SURVIVE THE VALLEY OF DEATH?

In his late work, *Capitalism, Socialism and Democracy*, he predicted the decline of the market economy.[16] He believed that monopolization and corporate bureaucrats would set capitalism on the path to self-destruction, while planned economies would prevail in the competition between systems. As we know today, history turned out differently. In the years of rebuilding after World War II, banks in Europe continued to provide for growth and innovation through loans, often secured by the state. In the United States, the government also directly and extensively promoted the new technologies—transistors, mainframe computers, semiconductors, communication technology, and space travel—being churned out by major corporations' R&D labs, foremost among them Bell Labs. And then came the disruptive innovation of venture capital, which used its wealth to clear the way for the digital revolution, first for the hardware, then for the software and the internet economy.

But since the financial crisis of 2008–2009 at the latest, the venture capital scene has largely moved away from its roots. Essentially, it tried to place market bets on proven technologies and business models, calculating that risk could be greatly reduced with sufficient diversification, while returns, with more and more capital in the asset class, would still increase. But risk-free risk capital is like squaring a circle. In 2021, it became clear that circles remain circles, even in venture capital—which is where a tremendous opportunity for the coming years can be found.

The next generation of venture capital investors will rediscover the curiosity that so successfully gave birth to so many new things in the second half of the twentieth century. The trend has already started to reverse. In the biotech revolution, the venture capital funding model is performing increasingly well, especially in the United States, because highly specialized venture capitalists and funds are once again combining vision and expertise. The same is true of the financing of quantum computers, nuclear fusion, and advanced battery technology. And with the large amounts of venture capital that have been flowing into climate technology since 2022 (and also thanks to government support), highly innovative founders will as a rule be able to avoid failing due to a lack of capital. In the United States, deep-tech investment gains an increasingly geostrategic touch. Investment company Andreessen Horowitz has founded a new investment practice called American Dynamism. This strategic unit supports founders and companies that can generate technological advancements in the national interest of the United States, "including but not limited to aerospace, defense, public safety, education, housing, supply chain, industrials and manufacturing."[17] The not-for-profit investment firm America's Frontier Fund operates based on a very similar logic. Chaired by former IBM CEO Sam Palmisano and Code for America Founder Jennifer Pahlka, America's Frontier Fund aims to ensure that the United States and its allies maintain a tech advantage amid a strong global power competition. The investment focus includes artificial intelligence, microelectronics, next-generation networks 5G and 6G, as well as synthetic biology, advanced manufacturing, new materials, and quantum sciences.[18]

In order for these positive trends to continue, it is also necessary to recognize and acknowledge venture capitalists' contribution to the innovation process. Why do we emphasize this? In popular narratives of disruptive innovation, genius founders are often the heroes of the story. If they have hatched their innovations with all their creative energies and then maximized many people's happiness, we won't insist on poking holes in their story. Meanwhile, among American progressives and the European mainstream, investors are often considered to be free riders who effortlessly make money with money. Investors are regarded as reminiscent of a casino owner overseeing the action behind a network of security cameras. In the end, the house always wins.

177

FINANCING INNOVATION LEAPS –
HOW CAN NEW TECHNOLOGIES SURVIVE THE VALLEY OF DEATH?

Narratives, including economic narratives, are often successful if they can be substantiated with memorable examples. For the narrative of the greedy and lazy investor, the history of capitalism offers numerous compelling examples. People also say that this type has been spotted in the venture capital scene. But maybe it's time to subject this narrative to some critical questioning. Venture capital democratizes capital by giving unknown and often inexperienced people with ideas and visions the chance to fail with their ideas and visions or—and this is the less likely possibility—the chance to succeed. Without a broad and permeable venture capital ecosystem, it would above all be men like Elon Musk or Bill Gates, Sergey Brin or Jeff Bezos, Jack Ma or (the controversial) Peter Thiel who have the opportunity to supply the farthest corners of the world with internet access from space, to set the research agenda for malaria, or to advance nuclear fusion, using their own resources and according to their own ideas. It is of course a welcome development if men of this sort also increasingly want to utilize their wealth according to the principles of impact investing, while using criteria for "improving the world" that the rest of us can understand. But at the same time, it would be disturbing and dangerous if they were able to monopolize disruptive innovation for themselves, like the ingenious villains or all-powerful corporations in dark cyberpunk novels like Neal Stephenson's *Snow Crash*.[19] To counteract this tendency, more, not fewer, venture capital investors are needed. This is not a new insight.

Without Johannes Fust, Johannes Gutenberg wouldn't have been able to set out on his book printing *aventur*. The print-based information revolution may have been postponed indefinitely. The investor Fust and the start-up founder Gutenberg had a bitter falling-out and became embroiled in legal battles even before their company's growth phase. People say that such things still happen today. Another thing that still seems to be the same for start-ups is that creative inventors have been dealt a bad hand at the "cap table," as the start-up and venture capital scene calls the distribution of shares and thus power among shareholders. In 1455, the court ordered Gutenberg to repay his financier 2,200 guilders because, it was claimed, he had engaged in a lucrative side business. The founder didn't have the money. Fust took over the company and forced out the first tech entrepreneur in history just after it had passed through innovation's valley of death and the printing press was entering the phase of profitable growth. (Incidentally, the same process is referred to today as the squeeze out.)

However, Johannes Gutenberg did not die a poor man, as is often claimed. Reboot. After his falling out with the clever businessman Fust, Gutenberg started over. He found new partners in a leading Mainz city official, Konrad Humery, and Archbishop Adolf II von Nassau, who provided advance financing for a new printing workshop. After his restart, the inventor Gutenberg also changed his attitude toward how he treated the proprietary knowledge that he had built up over decades. Up until the trial, he was very careful to preserve the secrecy of his intellectual property, his "secret sauce" for combining the press, movable type, and ink. After his entrepreneurial restart, Gutenberg shared his knowledge in new partnerships with printers who emulated and cooperated with him.

179

FINANCING INNOVATION LEAPS –
HOW CAN NEW TECHNOLOGIES SURVIVE THE VALLEY OF DEATH?

In his last creative period, the breakthrough innovator from Mainz also became a pioneer of forms of open collaboration that are to this day accelerating the further development and dissemination of innovative technologies. This open approach is the topic of chapter 6.

REINVENTING INNOVATION

WHAT CAN WE LEARN FROM OPEN SOURCE, OPEN DATA, AND OPEN INNOVATION?

"L-O-G-I-N"

On July 20, 1969, the world sat spellbound in front of their black and white televisions. Apollo 11 was safely orbiting the Moon. The *Eagle* lander touched down on the Moon's surface. Neil Armstrong climbed down the ladder and spoke his legendary sentence: "That's one small step for a man, one giant leap for mankind." The moon landing was the first technological breakthrough before a live television audience. And in addition to being an outstanding feat of engineering with many follow-up innovations in the ecosystem of American space travel, from magnetic resonance imaging and GPS to scratch-resistant sunglasses, it was of course also a PR blowout victory in the fight between Team USA and Team USSR. The world was deeply impressed by how America as a technology nation had planned the Apollo mission with military precision and executed it to perfection, from John F. Kennedy's visionary speech in 1961 to the landing of the capsule in the Pacific Ocean south of Johnston Atoll. Beaming with pride in its space race accomplishment, the American public unfortunately missed the real innovative feat of 1969.

One hundred and three days after Armstrong left his footprints in the Moon's dust, on October 29 at 10:30 p.m., UCLA computer science student Charley Kline typed three letters on the keyboard of an SDS Sigma mainframe computer: l - o - g. With a slight case of nerves, Professor Leonard Kleinrock looked over his student's shoulder. In contrast to the rest of the world, it was clear to Kleinrock, Kline, and a few other young scientists in the room that if the experiment succeeded, it could spark a new information revolution. UCLA's SDS Sigma was connected to another mainframe, Stanford Research Institute's SDS 940, via telephone line. The goal was for the computer in Los Angeles to log into the one in Silicon Valley. The message "l - o - g" from Kleinrock's team was then supposed to be automatically completed with "i - n" by the Stanford computer. Los

183

OPEN SOURCE, OPEN DATA, OPEN INNOVATION –
CAN WE REINVENT THE WORLD TOGETHER?

Angeles typed "l." It arrived. Then "o" arrived. Unfortunately, then the SDS 940 in Northern California crashed. The first message sent over a communications network that would later become the internet was the mangled sequence "lo." It didn't provide much material for a PR sensation.[1]

It's no coincidence that the greatest innovative leap in our lifetime, and the greatest innovative breakthrough for information since Gutenberg, tiptoed so quietly into the world. Even today it's difficult to visualize what the internet actually is. It has deeply permeated our lives, our work, and our society, probably more deeply than even the most optimistic researchers at UCLA and Stanford could have imagined that evening. Today, the internet is always on and everywhere, or "ubiquitous," as the computer visionary Marc Weiser called it in his legendary essay "The Computer for the 21st Century."[2] But we only see it superficially in the form of pixels on our various screens. Or not at all, when it's humming away in the background to control logistics systems, value creation processes, or our lives. The internet has become as powerful as it is mundane. In the process, we've forgotten why it was a much more unlikely innovative leap than the automobile, the light bulb, or antibiotics, and that its origin was the exact opposite of the technologically brilliant achievement of the Moon landing.

It's true that the internet received start-up funding from the military via (D)ARPA (didn't everything?). The military was interested in building a highly robust communications network that would still be operational after a potential Soviet nuclear strike. But then its development took on a life of its own, and ARPA was farsighted enough not to make any serious attempt to reclaim control of ARPANET as a closed project. It was already clear that taking a systematically closed approach to networking the world would be a contradiction in terms. The internet and the World Wide Web built on top of it became a triumph of cultural and technological openness. It is the improbable team accomplishment of great minds like Vint Cerf, Robert Kahn, and, later, Tim Berners-Lee, as well as a large community of developers whose names never became famous but who were committed to three principles.

1. Open protocols and open standards: the technological foundations of the internet, especially the data transmission protocols, are openly documented and accessible as source code.
2. Decentralized structure: Whoever wants to become part of the network with a new node (server) doesn't have to ask anyone for permission. Once the node meets the technological requirements, the network expands to include it.
3. Freedom from licensing restrictions: Anyone can use the protocols to join the full network. Anyone can check the protocols (or "hack" them) and continue to develop them. They're free to use.

These three principles are the basis of the information revolution of the past five decades.

Around an hour after the "lo" message, the login finally worked, and the first small data packets were sent. Three weeks later, the teams set up a permanent connection—and the first two "nodes" of ARPANET were born. In December, the network had four nodes. By summer 1970, it stretched to the East Coast and the University of Hawaii. The Hawaiian node connected to a radio network that spanned several islands. In 1971, the first e-mail was sent. In 1973, Norway became the first European country to link in via satellite, and Great Britain soon linked in via submarine cable. Universities became the drivers of development. In Canada, Hong Kong,

185

OPEN SOURCE, OPEN DATA, OPEN INNOVATION –
CAN WE REINVENT THE WORLD TOGETHER?

1969	1971	1973
ARPANET connects the first mainframe computers	The first email is sent	Development of the TCP network protocol (later TCP/IP)

	1979	1974
	Development of a civil Internet governance structure (initially the ICCB)	First commercial data transfer service

1983	1988	1989
Military ARPANET finally splits off (MILNET)	"Morris worm" paralyzes one in ten Internet servers	Hypertext protocol for the WWW developed / First commercial Internet services

	1994	1993
	Founding of Amazon, beginning of the dot-com boom	The "Mosaic" browser makes the network accessible to everyone

1996	2000	2002
Google founders start their search engine "BackRub"	Dot-com bubble bursts	Breakthrough of mobile Internet thanks to 3G (UMTS)

	2007	2004
	iPhone 1 comes to market	Founding of Facebook, rise of social media

2009	2013	2023
WhatsApp goes online	Samsung makes first 5G data transfer in the lab	More than 5.2 billion people use the Internet

Australia, and continental Europe, they established their own nodes. New transmission protocols, later designated with the collective term TCP/IP, led to data exchange that continuously increased in speed, reliability, and volume. Through various associations such as NSFNET, ISOC, and ICANN, the internet community created a semiformal administration for itself, including the scuffles and disputes typical of grassroots self-administration. At the beginning of the 1980s, the military side of ARPANET formally split off, and a year later the internet got its phone book in the form of the Domain Name System. In 1988, the first malware, the "Morris worm," spread and paralyzed around 6,000 of the 60,000 servers that existed at the time.

At the international nuclear research facility CERN in Switzerland, the British physicist and computer scientist Tim Berners-Lee was already working on a new, open transmission protocol that everyone could use free of charge: HTTP. Berners-Lee's hypertext transport protocol led to browsers. The browsers turned the internet of scientists into the internet for everyone, the World Wide Web. With the Web, the open, decentralized, license-free community of network developers had proven once and for all that the essence of digital innovation is openness. Cooperation is more powerful than competition. Open collaboration created the basis for the development of all the digital systems that have so dramatically changed our work, our society, and our lives over the past three decades. What we overlooked for a long time, though, is that open innovation laid the foundation for the platforms and Big Tech companies that for around fifteen years have been increasingly monopolizing the leading programs and applications of the networked world—and technologically closing them for their own benefit—in a dialectical process practically drawn from a Hegelian textbook.

THE ESSENCE OF DIGITAL INNOVATION IS OPENNESS.

187

OPEN SOURCE, OPEN DATA, OPEN INNOVATION —
CAN WE REINVENT THE WORLD TOGETHER?

OPENNESS, POWER, AUTHORIZATION

It's time for a moment of full disclosure again. In this chapter, we're far from being unbiased observers. As a programmer and entrepreneur, Rafael produced open-source software for three decades, motivated by the desire to drive technological change with free software and at the same time to preserve the open character of the internet. In 2018, Thomas and Oxford professor Viktor Mayer-Schönberger initiated a worldwide debate about opening up access rights to data with their book *Reinventing Capitalism in the Age of Big Data*.[3] Their central thesis was that data monopolies concentrate innovation in a few hands and thus slow it down, so they need to be legally compelled to share data with others.

But precisely because of our biographical connection to openness in technology and machine-readable information; our many discussions of the advantages and challenges of open-source software and the technological, legal, and cultural hurdles faced by data sharing; the sometimes legitimate and sometimes lobbyist-organized attacks on our positions—for all these reasons, in this chapter, we would again like to discuss the following question: What degree of openness in a given context increases the chance of at least accelerating innovation or, hopefully, facilitating major innovative leaps?

In the search for answers, three aspects seem particularly relevant to us: (1) open versus proprietary software or technology, (2) open data access versus exclusive use of data, and (3) open versus restrictive treatment of patents and intellectual property.

OPEN

TRALIZ

LICENS

DECEN -
ED
E - FREE

$\rightarrow$ **1.**

Open Software Development as an (Inexpensive)
Driver of Innovation

"Software is eating the world."[4] The sentence comes from Marc Andreessen. He knows what he's talking about. Today, Andreessen is one of the most successful venture capitalists in the world. At the beginning of the 1990s, he wrote the code for the Mosaic browser. Based on this, he later founded Netscape, the first internet software company, which helped to make the wealth of information on the internet easily accessible to everyone on the World Wide Web. "Software is eating the world" became the catchphrase of programmers, not only in Silicon Valley but worldwide. Even non-techies can sense that this is the case. But what's gotten less attention is the next step: "Open-source software is eating software"—and it has a voracious appetite.

The story of Linux as an open-source operating system for PCs has been frequently recited, often with some ambivalence: it's all very nice, the penguin logo is really cute, but ultimately Linux is something made by nerds for nerds. This ambivalence has contributed to open-source software being chronically underestimated in terms of its technological importance and contribution to global value creation. The data are anything but ambivalent, however.

While the profits of Microsoft, Apple, SAP, and other major tech firms are increasing surprisingly rapidly, the use of open-source software is growing significantly faster than the major software corporations' proprietary systems. In other words, Big Tech is actually losing market share to open-source year after year. According to a report by the US tech services company Synopsys, 99 percent of software in 2020 contained at least some open-source code. Perhaps even more impressive is that 70 percent of the software used worldwide was

191

OPEN SOURCE, OPEN DATA, OPEN INNOVATION –
CAN WE REINVENT THE WORLD TOGETHER?

OPEN-SOURCE SOFTWARE IS EATING SOFT-WARE — AND IT HAS A VORA-CIOUS APPETITE.

essentially developed on the basis of open source. That's twice as high as in 2015.[5]

We're also experiencing a notable turning of the tide in enterprise software, especially in the United States and Asia. Even in Europe, more than half of all companies are modernizing their IT infrastructures with open-source technology.

This paradigm shift in enterprise IT systems is being spearheaded by Enterprise Open Source (EOS), as it's known. This is software that's created using the collective approach of open software development and whose functions are freely available, including human-readable source code. Companies such as RedHat and SUSE Linux then make EOS palatable for corporations and governments for a license-like fee. In return, these companies provide help with installing, customizing, and maintaining the software and—most importantly—a telephone number that a competent employee will answer immediately. We're also seeing a strong global trend toward free software in the public sector, with governmental organizations and public administration in the United States leading the way. More and more often, what you hear there is "public money, public code."[6]

The reasons for switching to open source are the same in the private and public sectors: open-source software is no longer designed by nerds for nerds. Instead, over the past ten years, it's become much more convenient to use. It offers more flexibility and leaves users less dependent on a single provider. And it provides a foundation for its own continued development, since customers can use all preceding work for their own purposes without restriction: for learning to use the software, for expanding the program, for improving the code. Open source creates systems that work well together, among other things thanks to open standards and open interfaces, and at the same time it creates transparency for users. Of course, free software is never truly free of charge. Someone must install, configure, and maintain it, and that at least costs time, and usually money for external assistance as well. But it's probably not just a myth told by software developers that the full cost of open-source software is significantly lower on average than for proprietary systems, whether it's installed in a company's offices or, increasingly, hosted in the cloud.

More user-friendliness for nonexperts, fewer lock-in effects, high transparency and trustworthiness, and low costs, however, don't explain why even the Big Tech companies that have become rich and powerful based on proprietary systems are increasingly relying on open source. Microsoft, Amazon, Google, Facebook, IBM, Oracle, Salesforce, SAP, and other major firms are increasingly building open-source software into their own products and selectively supporting developer communities when it promotes their interests.

What may seem like an astonishing paradox at first glance is soon revealed as a logical consequence on closer inspection. The smart people in IT at Big Tech companies know the technological history of the internet. They know that open development in a digital context is far superior to proprietary systems at producing radically better solutions. Does this sound like we're preaching the open-source gospel? In this case, it's not the data that provide evidence, but a look at the technologies behind the major IT trends.

The power of open source is at work in the background of all the important IT innovations of the last few decades. Without open-source development, there would be no high-performance databases and no cloud. The major advances in machine learning have been achieved through freely usable algorithms that AI researchers can download free of charge. Researchers will also find numerous free applications online. Cryptocurrencies and the blockchain technology behind them are open-source development in its pure form, decentralized in structure and unrestricted. The Big Data revolution was made possible by

193

OPEN SOURCE, OPEN DATA, OPEN INNOVATION –
CAN WE REINVENT THE WORLD TOGETHER?

OPEN-SOURCE TECHNOLOGY
BEHIND MAJOR IT TRENDS

WEB DEVELOPMENT

PHP, Angular JS,
Node js, Eclipse
Che, React, …

**INFRASTRUCTURE
SOFTWARE**

Linux, Xen, KVM,
Ceph, OpenStack

BIG DATA

Apache Hadoop, Apa-
che Spark, Apache
Cassandra, Mongo DB

MOBILE

Android, Apache
Cordoba

MACHINE LEARNING

TensorFlow, Apache
MXNet, PyTorch

DEV OPS

Jenkins, Chef, Puppet,
Ansible, Terraform,
Openshift, Cloud
Foundry, Docker

Source: primcore.com, Why Open Source

OPEN-
SOURCE
SOFTWARE
IS NO
LONGER
DESIGNED
BY NERDS
FOR NERDS.

open-source software like Hadoop and Mon-
goDB, and vice versa. The world's most widely
used data analysis software, R and Python, are
freely available online thanks to their highly
committed army of volunteer developers. With-
out open source there would be no Android (and it would be nice if
the people at Google remembered that a little more often). There's so
much open-source code in self-driving cars that if you printed it out
in a medium-sized font, it could circle the Earth three times. And the
weather forecast would be even less reliable without open-source forecasting algorithms, as would almost all digital forecasting applications.

The story of the rise of the internet is repeating itself in many digital technologies, applications, and industries: open software development is the primary engine of IT innovation. The more clearly we recognize this, the more horsepower this engine can put out. And maybe
this time we'll avoid having a few superstar firms monopolizing the
most innovative applications for themselves and closing off access step
by step to formerly open technology for their own benefit. This systematic takeover has been going on for years. Microsoft's acquisition
of GitHub, the largest platform for managing open-source projects, is
just the most prominent example.[7] You don't have to be a dyed-in-the-
wool open-source enthusiast (like Rafael is) to come to the conclusion
that as a rule, Big Tech companies give little and take a lot when they
participate in open-source projects. Again and again, they instead with-
hold their innovative energy from open development. When they get
involved, sooner or later the source code is no longer open, the license
is no longer free, or the protocol is no longer open, decentralized, and
unrestricted. Then the creative energies of many people are no longer
able to contribute. At that point, creative destruction benefits only a few
people: Big Tech shareholders. That's certainly not what the inventor
of the internet had in mind.

195

OPEN SOURCE, OPEN DATA, OPEN INNOVATION –
CAN WE REINVENT THE WORLD TOGETHER?

$\rightarrow$ 2.

Open Access to Data and Information as a Resource for Innovation

In fall 1910, a young Austrian economist accepted his first professorship in Czernowitz, on the northeastern tip of the Habsburg Empire.[8] The ambitious Viennese professor lost no time in assigning a lengthy reading list to his students. But to the professor's surprise, the students at the Franz Josephs Universität (today Chernivtsi University in Ukraine) were unable to do their homework. A librarian refused to lend them the books, even though they were available in sufficient numbers for everyone. The young professor challenged the librarian to a fencing duel. The professor's name was, possibly not as much of a surprise, Joseph Schumpeter. The librarian left the fencing arena with a deep cut in his shoulder and from then on granted access to the books. If the impulsive Schumpeter, the intellectual predecessor and hero of disruptive innovators everywhere, knew about access rights to information in the age of data, he would probably challenge numerous companies, privacy activists, and regulators to a duel, one after the other.

The most important source of innovation is access to knowledge. Knowledge arises through the contextualization of information. The most important information today is machine-readable. We refer to machine-readable information as data. But the age of data has brought about an interesting paradox. On the one hand, we're overwhelmed by a flood of data. Individuals, organizations, and governments can't cope with the flood of digital information. Instead, they give up. On the other hand, many people and organizations that could drive innovation forward don't have access to the data they need. There are three essential reasons they're starving for data in the middle of a data flood.[9]

1. The big commercial players and their platforms have succeeded in creating an environment where they can use data exclusively for their own purposes—including data that arise only through interaction with others. These "others" are both people and machines that belong to others.
2. Public institutions have been unwilling or unable to make governmental data, administrative data, or the data of semigovernmental organizations available on a large scale. To this day, open data largely remain a distant hope.
3. Much of the data that are valuable for innovation, especially in the life sciences and digital education, isn't being recorded at all because there's no immediate commercial interest in it yet. Or the data are lying inert, recorded in different standards, and fragmented across different databases. Often data protection requirements must be met in order to make data usable. This is usually more of a technological hurdle than a legal obstacle, but unfortunately the technological hurdle is often quite high.

In all three areas, there's reason to hope that things will improve, even without dramatic intervention. Regulators in Europe, the Biden administration, and also China's digital bureaucrats have recognized that letting Big Tech companies monopolize the data on their servers is good for the digital machinery of power and those who operate it but bad for innovation. As often in digital regulation, the EU Commission is spearheading the action. The European Data Act, which is currently under negotiation, seeks to compel the data giants to make some of their data accessible to others.[10] This can be seen as learning from the pandemic, where scientific data sharing has sped up vaccine development while many data troves of Big Tech companies, that could have been of

197

OPEN SOURCE, OPEN DATA, OPEN INNOVATION –
CAN WE REINVENT THE WORLD TOGETHER?

value for public health measures, had not been accessible for the public good. When new data access rules are in debate, the army of Big Tech lobbyists often makes noise about "expropriation" and "confiscation." That's incorrect.

There is no right per se to own data, especially not when the data have been created through interacting with others. The usefulness of data is created by their use. In economic terms, data are also a nonrival good. They can be repeatedly used in many ways without losing quality or value, just as a novel becomes no less exciting if many people read it. It is therefore possible as a matter of law, common sense as a matter of economics, and necessary for society to open access to essential data wherever data oligopolies and monopolies restrain progress—even if these monopolies are skilled at presenting themselves as especially innovative companies.[11] In Europe, it would only require an addition to the General Data Protection Regulation: a General Data Usage Regulation. Monopolies have never been good for innovation. Data monopolies are acts of theft against progress, and the debate about it in our society has only just begun.[12]

In addition, we have the impression that the coronavirus pandemic has finally breathed new life into the bitter debate about open data. It has become clear to everyone involved in the debate that government at long last needs a digital reset. Open government data are capable of much more than creating innovative applications for smart cities—as they have done in Barcelona. With open data governments can significantly increase their digital IQ. The Scandinavian countries and Estonia, the digital star pupil, demonstrate how e-government, the competent use of technology for governance, and a culture of open data go hand in hand. Governments can and must be the role models here, because their data are by definition a public good. Making this resource artificially scarce and inaccessible to those who could use it as a basis for innovation is not only economically stupid. It also erodes belief in the future viability of one's own country and therefore endangers democracy.

We see the greatest potential, however, in a more intensive and more intelligent use of valuable data for which open access is currently restrained neither by monopoly power structures nor by a lack of goodwill among those involved. Everywhere we look, research projects and commercial initiatives are making health data useful for radically better diagnostics and treatments, with the Human Genome Project serving as a forerunner for the age of data. In that project, researchers decoded the human genome in open-source style and laid the foundation for rapid advances in "red" (medical) genetic engineering. Breaking down data silos and standardizing health data to make them useful is a tedious process.

As already indicated, data protection requirements are an important and difficult topic, especially when it comes to sensitive health data. The many active players in the broad field of health innovation have different views as to the level of technological requirements for pseudonymization, anonymization, and encryption of health data for researchers to gain access to them. The growing consensus, however, is that we must do everything possible to make as much data as possible available to as many researchers as possible, while ensuring security and respecting personal privacy, to finally defeat the great scourges of humanity, from cancer to Parkinson's disease to dementia. The good news is that with consensus on data exchange, the technological possibilities are also growing to overcome the obstacles related to data protection. New cryptographic methods are helping. The ideal solutions are databases in which, thanks to these new cryptographic methods, researchers can make full use of health data containing personal information without having access to the actual personal characteristics. The personal details remain concealed behind a technological veil. This procedure can also be applied to many other areas, and once the

199

OPEN SOURCE, OPEN DATA, OPEN INNOVATION –
CAN WE REINVENT THE WORLD TOGETHER?

technological problem has been solved, nothing stands in the way of open access to research data except human narrow-mindedness and greed. We will only be able to break down silos in the data structures when we overcome the silo mentality in our minds. Perhaps the most impressive example of the boost to innovation through availability of data is—once again—a DARPA project. When military GPS data became available in May 2000, it sparked the navigation revolution that today tells us in real time, among other things, that it's faster to take the county road right now rather than the interstate.

→ 3.

Patent Protection, Patent Deception

The recipes in a cookbook are algorithms used by people. They guide us sequentially, step by step, so that in the end we obtain the desired result. Computer algorithms are fundamentally excluded from patent protection because they're based on mathematical methods. Algorithms are mathematics, not technology, and math doesn't enjoy patent protection. At least for recipes, the Greeks of the sixth century BCE saw things differently.[13] "If one of the cooks invents a new, delicious dish, no one but the inventor himself will be allowed to make use of this invention until a year has passed. During this time, he will have the commercial profit from it, so that the others exert themselves and try as rivals to outdo each other in such inventions," reports the historian Athenaios about a decree in the Greek colony Sybaris, in what is today the Italian region of Calabria. These two sentences, written around 2,500 years ago, capture the essence of the modern understanding of patent protection. The inventor is allowed to forbid others to copy his invention. The prerequisites for this are novelty and commercial use. The proprietary right is for a limited time. And the Sybarites had even already grasped the primary justification for modern patent law: exclusive use and the resulting increased benefit of use promote the spirit of invention.

Do we really need patents? And do only proprietary rights and the prospect of monopoly profits for a limited time encourage innovators and their financiers to put in the advance work for an invention? Or are patents an innovation-hostile dead end in the history of law and technology? Do they hinder new technology from rapidly developing and spreading? This dispute has been ongoing ever since patent law began taking concrete form in Europe, step by step and country by country, in the fifteenth century. The dispute became especially bitter in England in the course of the Industrial Revolution. The dispute will likely continue as long as there are patents. For us, the default is always openness. But with patents, the only answer that seems to make sense is legal experts' standard answer to all legal questions: it depends. When President Biden declared in May 2021 that he would support the lifting of patent protections for coronavirus vaccines as part of World Trade Organization negotiations, it delighted the US Left and shocked the American pharmaceutical lobby.[14] Germany too was immediately split into two camps. That Biden's statement made political waves in Germany is understandable in two respects. Finally, there is another innovative leap (mRNA technology) coming out of Mainz, and *now* you want it to be the one exception where world socialism applies to intellectual property?

Experts without ideological baggage, on the other hand, quickly agreed that suspending the patents of Moderna, BioNTech/Pfizer, and CureVac wouldn't get a single dose any faster to those countries that needed the vaccine so urgently. The problem wasn't the patent, because licenses can easily be granted at a fair price, and maybe a little political pressure would help with that. The real bottleneck

201

OPEN SOURCE, OPEN DATA, OPEN INNOVATION —
CAN WE REINVENT THE WORLD TOGETHER?

was production. If you wanted to vaccinate the world quickly against COVID-19, you would have to ramp up production and set aside the vaccine nationalism.

In retrospect it seems obvious to us that the vaccine patents were precisely the wrong example for discussing the necessity of simplifying and reforming patent law. Considering its pretax profits, our sympathy for the pharmaceutical industry is also limited. The duration of proprietary rights and the design of mandatory licensing rules similar to fair use can and should be discussed. But the extremely high development effort, the high risk of failure in testing phases, the bureaucracy overseeing regulatory approval, the strict liability rules, and the easy copying of the product provide sufficient reason for the assumption that no large pharmaceutical company, no venture capitalist for biotech start-ups, and no private equity manager would invest billions in the development of new vaccines and medications in the future without a secure legal framework for exclusive exploitation for a limited time. From the innovator's point of view, vaccines have in addition to all these risks the complicating factor that when a vaccine finally arrives, no one may want it anymore. This is what happened to the developers of the Ebola vaccine during the 2014 outbreak in West Africa. When a promising compound entered the third phase of clinical trials in 2015, the epidemic was under control, and it became extremely difficult for the developers to recruit volunteers as test subjects.[15] But compared to other industries and technologies, this tends to be the exception rather than the rule.

If we look at the history of innovation over the past two hundred years, two developments stand out. First, the number and severity of patent protection regulations have continued to increase—and along with these regulations, so have the bureaucratic requirements and costs of securing a patent. Second, however, this hasn't prevented innovators from investing more and more time, money, and energy in establishing

THE MOST IMPORTANT SOURCE OF INNOVA- TION IS ACCESS TO KNOWL- EDGE.

202

CHAPTER 6

their own patent power, only to then defend it in bitter legal feuds involving armies of expensive patent attorneys or, even worse, using it destructively against others. In the educational and entertaining book *How Innovation Works*, British politician, zoologist, and entrepreneur Matt Ridley tells a series of anecdotes about innovators who sacrificed many of their creative years on patent disputes instead of focusing on their innovations.[16] For Ridley, James Watt and Guglielmo Marconi, Thomas Edison and Nikola Tesla, and the Wright brothers are not only heroes of innovation but also masters of wasting their creative energy. Watt wished for a "noose around the neck" of his competitor Richard Trevithick. Edison and Tesla waged a war between AC and DC electricity. Marconi fought against Telefunken's SOS emergency signal for years. In the years leading up to World War I, the US aircraft industry stagnated because its pioneers kept suing each another. Meanwhile, French innovators cooperated with each other and flew past the Americans technologically. The history of technology has seen more than its share of repeating loops. In 1984, the Semiconductor Chip Protection Act established particularly tough patent rules for high tech in the United States. What followed were the patent wars of the chip manufacturers in the 1990s, which were followed by combat between smartphone providers in the 2000s. Innovation in chips and high-end hardware has meanwhile migrated to South Korea, Taiwan, and China. Moderna starting a patent rumble with BioNTech/Pfizer over COVID vaccines, unfortunately, adds the next chapter to this never-ending story.[17]

In his book *Launching the Innovation Renaissance*, Canadian American economist Alex Tabarrok presents an interesting hypothesis based on the question, "Do patents help technological progress?" [18] Tabarrok draws a parallel to the Laffer curve familiar from

203

OPEN SOURCE, OPEN DATA, OPEN INNOVATION –
CAN WE REINVENT THE WORLD TOGETHER?

debates about taxation. At some point, higher tax rates lead to lower tax revenues. At some point, patent protection doesn't promote innovation but slows it down, because intellectual property hinders the further development and dissemination of technology more than it motivates innovators to develop fundamentally new things. And we would like to add that at some point, patent bureaucracy no longer ensures legal security, but legal blockades and rising legal fees instead.

Exactly when these tipping points are reached—where the benefits of patents are outweighed by their disadvantages for particular technologies—would be an extremely interesting research topic. Even Alex Tabarrok can't precisely identify these points. We would be very surprised if we hadn't passed these tipping points many years ago. A pragmatic, nonideological reform of patent law, as also suggested by Tabarrok, would be the logical consequence of this situation. The key points would be faster, easier, and less expensive patent approval; shortening of patent durations; and rapid arbitration with elements of mediation for fair use licenses.

This kind of reform would do innovation a great service without completely undermining the patent system. Proprietary rights to technology are a balancing act. They deliberately slow the dissemination and further development of new technologies in order to create an environment conducive to creating new technologies in the first place. It's necessary to restore this balance in favor of the possibilities of open cooperation. The development of the COVID-19 vaccines in the spirit of "coopetition," or a healthy combination of competition and cooperation, could point to the way forward. Property rights and openness are not mutually exclusive.

The rapid development of effective and safe vaccines was possible because researchers around the world have worked together with exemplary openness. The world owes the most effective vaccines to the prior work of the innovators behind the mRNA platforms. The most effective vaccines were then further developed in an environment of fierce competition with the prospect of high profits. It worked out well.

OPENING UP OPEN INNOVATION

When someone coins a phrase that instantly captures the essence of what had been a vague development or that perfectly describes the current zeitgeist, the phrase can spread as fast as a good meme on Instagram. Sometimes it grants the concept's inventor fifteen minutes of fame, and in some cases it's even possible to build an academic career on this one act of word creation. To ensure that your claim to the concept is undisputed, it's helpful to make it the title of a book. Sometimes all it takes for a new concept is to combine two common words. In 2003, a previously little-known lecturer at Harvard Business School pulled off this trick. His name, Henry Chesbrough. His term "open innovation" perfectly captured the managerial zeitgeist of the time.[19]

For several years, Chesbrough, an economist, had been researching how the networking of the world with digital technology was changing innovation in large companies. To do this, he took a close look at the microchip and computer industry. His initial thesis was that people who bring about technological paradigm shifts would provide particularly early examples of how the innovation model was changing. And lo and behold, Chesbrough found ample evidence that the internet was actually living up to its promise in the high-tech industry. Digital technology creates the opportunity for better networking across labs or corporate campuses and to collaborate more openly with creative minds two blocks away or on the other side of the world. Companies that take advantage of this opportunity are demonstrably more inno-

205

OPEN SOURCE, OPEN DATA, OPEN INNOVATION –
CAN WE REINVENT THE WORLD TOGETHER?

THE INNOVATION LAFFER CURVE

Incentives based on
patent protection

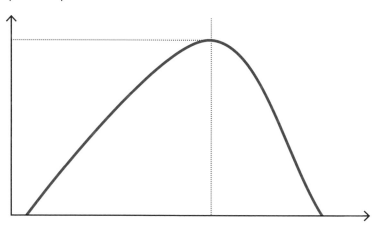

Slowed dissemination
due to
proprietary rights

vative than closed research and development departments, even those with a lot of very bright people working in them.

Practically overnight, the book *Open Innovation* became mandatory reading for corporate innovation managers worldwide. Since the mid-2000s, corporate Goliaths in every industry have been seeking greater openness by sponsoring competitions for new ideas, systematically integrating lead customers and suppliers into innovation processes, and strategically building innovation ecosystems with smaller partners. In some cases it worked well (particularly in the pharmaceutical and chemical industries with open platforms like InnoCentive[20]) and in some cases less well (as with automobile and consumer goods manufacturers). Striving for openness was certainly always useful just for the sake of the learning experience. And it was at heart a logical consequence of sociotechnological developments. The managerial zeitgeist always follows technological paradigm shifts that entail social change. Companies that refuse to change will sooner or later disappear from the market. That too is the nature of innovation.

On the basis of the term "open innovation," Henry Chesbrough was able to reinvent himself. Propelled by his successful publication, he became a professor at the University of California, Berkeley, and rose into the major league of academically based management consultants. His success was certainly well deserved. Sadly, however, the term "open innovation" has remained trapped in the world of management consulting since 2003 alongside a cast of phrases like "change management," "customer centricity," or "new work." But the term "open innovation" was born for greater things, and we should open it up.

For us, open innovation combines the collaborative principles of open-source software

OPEN INNOVATION IS BASED ON A COLLABORATIVE CULTURE OF OPEN COLLABORATION IN WHICH THE INNOVATORS GENEROUSLY SHARE THEIR INFORMATION.

207

OPEN SOURCE, OPEN DATA, OPEN INNOVATION –
CAN WE REINVENT THE WORLD TOGETHER?

development with a willingness to make information much more accessible than is common today. We understand that information asymmetries can lead to competitive advantages, and that this is of course an element of the Schumpeterian innovation game. But in the age of machine-readable information, a few players armed with strategic foresight and a bag of dirty tricks are increasingly blocking access to one of the most important resources for innovation in the twenty-first century. Data monopolies enable these few firms to generate monopoly returns at the expense of progress. Open innovation encompasses a basic political understanding that data have the character of a public good. Governments need to lead the way with an open data culture. Open innovation also means bringing the costs and benefits of patents back into a sensible balance and promoting the use of patents on fair terms. To put it in one sentence: open innovation is based on a culture of open collaboration in which the innovators generously share their information, and the entrepreneurial risk of the inventors and developers is adequately safeguarded without unnecessarily hindering continued development.

Does that sound like an airy vision? This vision was already the lived reality of daily experience in a major innovation ecosystem long before the term "open innovation" became a corporate buzzword. This ecosystem was created by an unorthodox but farsighted governmental intervention: the announcement of the partial break-up of AT&T in 1953.[21]

With technical expertise, corporate strategic skills, financial power, and the occasional dirty trick, the American Telephone & Telegraph Company had seized a monopoly position for itself in the American telephone market since the 1880s. Since 1913, AT&T has been locked in an ongoing legal feud with fair trade regulators in Washington. The key question was, Do the network effects of a telephone network automatically lead to a natural monopoly? The more participants a network has, the better it is for everyone. Or does even a telephone network need competition in order to make the service affordable and innovative? Europe opted for state monopolies. The United States took a more innovative path.

In 1956, AT&T attorneys negotiated an interesting compromise with the Justice Department. The company was allowed to retain its monopoly on telephone services and thus continue to generate enormous profits. In return, however, the corporation's giant research and development center, the legendary Bell Labs, had to open its existing patents free of charge to all American companies. The labs had to make all future patents available for a modest license fee. To understand the effect of this change, it is necessary to take a look at the breadth of the patent portfolio. Bell Labs had well earned its nickname as the "Idea Factory." From transistors and microprocessors to vacuum tubes, lasers, coaxial and fiber optic cables to solar cells, in the 1950s, there was virtually no interesting high-tech field in which AT&T's idea factory wasn't at the head of the pack in terms of knowledge and patents.

Internally, even before the patents were opened, Bell Labs were known for a research and development culture that today would easily fall under the "open innovation" label. Bell Labs helped invent the culture of intensive interaction with universities and geeks of all stripes, tolerance for mistakes, flat hierarchies, and avoidance of the not invented here syndrome. The legally compelled opening of the patents led to an even more radical opening of the organization. Now there wasn't even an economic reason to hoard knowledge as a way to dominate competitors. The unconventional compromise became an important impetus behind the rise of Silicon Valley to its status as the world's ideas factory. Open patents played an important role. But above all, the legal compromise initiated a cultural paradigm shift toward open collaboration in an innovation ecosystem. This was the truly revolutionary effect of the antitrust compromise of 1956. It ultimately made

209

OPEN SOURCE, OPEN DATA, OPEN INNOVATION –
CAN WE REINVENT THE WORLD TOGETHER?

it possible for an initially loose association of breakthrough innovators to send the message "lo" half a century ago, and then jointly develop an open and collaborative space called the internet.

If "open innovation" can find its way back to its roots, it will help bring about great leaps in progress for many new innovations in the next fifty years.

TECHNO-OPTIMISM

HOW FAR IS IT TO UTOPIA?

THE HUMMINGBIRD EFFECT

Once again, nature shows us how it's done. And Charles Darwin pointed out its precise method. Sometime in the Cretaceous period, more than 66 million years ago, flowers began using bright colors and scents to signal to insects: get all the pollen you want right here! The Cretaceous insects had no more need to understand the evolutionary purpose of this than bees do today. It was enough for them to fertilize the next flower they flew to. Over many generations, the insects evolved better forms and tools for gathering the pollen more efficiently. The plants in turn enriched their blossoms with nectar, which incited evolution's random gene mutations to equip insects with better sensors to help them find the best flowers with the juiciest nectar. Biologists refer to this form of evolutionary partnership as coevolution. Darwin devoted an entire book to orchid pollination. And he also observed that coevolution can go in surprising directions. With the evolution of flowers and nectar for insects, the surprises include a bird species that is extremely small, has a particularly fine, pointed beak, and has flight abilities that other birds can only dream of. Despite its traditional avian skeletal structure, a hummingbird can hover in the air over a flower and suck nectar like a bee. Evolution taught it to let its wings rotate.

The American author Steven Johnson uses this example from evolutionary biology in his book *How We Got to Now* as a metaphor for similar, recurring, and significant spurts in technological innovation.[1] The "hummingbird effect" is his term for surprising innovations caused by innovations in other fields. But in contrast to chaos theory's well-known butterfly effect, the causalities in cascades of hummingbird-like technological evolution are quite visible. Johannes Gutenberg's innovative leap in printing created a market for books and instigated an information revolution (as described in chapter 5) that made the Enlightenment possible. Because of all the books, there were more people who learned to read. They increasingly found that they were farsighted and needed glasses. The need for vision aids in turn led inventors to study optical lenses. One innovation that followed from this was as obvious as it was world-changing—the microscope. Without microscopes, we still wouldn't know today that our body is composed of cells. And a young German doctoral student, Ingmar Hoerr, wouldn't have recognized in 2001 how mRNA in cells can ward off the attacks of deadly viruses; a finding on which Katalin Karikó and Drew Weisman could build to find out how to apply nonimmunogenic, nucleoside modified RNA, a technology that was finally licensed by both BioNTech and Moderna to make mRNA vaccines against COVID.

Scientists often rightly criticize popular science publications for being overly fond of metaphors. In this book, we've reduced complexity at some points by using figures of speech and anecdotes. Perhaps we've simplified some things too much. But the metaphor of the hummingbird effect does seem very useful to us when we look from the present to the technological future. Because in many fields of technology, there is currently a high probability of a whole flock of hummingbirds, so to speak. Or to put it less figuratively, there are many reasons to be optimistic about technology, because highly original follow-up innovations in the coming decades will help us to meet the major challenges that today often seem distressing, dangerous, and unsolvable.

TOO CHEAP TO METER

Let's start with the energy problem. Today we mostly discuss energy with a focus on reducing CO_2 emissions and climate change. On the one hand, this makes perfect sense, but on the other hand, this perspective is far too limited. In heated discussions focused on the minutiae of the true environmental impact of electric cars or the economic costs of CO_2 emission certificates, we often overlook one thing: If we solve the energy problem, if we make clean energy available anywhere in the world at an affordable price, a whole host of additional global problems will also vanish. Not only could we get control of climate change, but there would also be fewer wars, less poverty, and less migration. In the 1960s, techno-utopians and fans of nuclear energy started describing this vision as the "too cheap to meter" scenario. With nuclear fission alone the plan hasn't panned out, yet.

But what if clean energy were so inexpensive everywhere in the world that instead of consumers being billed for how much they used they would be asked to pay a small, flat-rate fee, like we do with the internet? Each country would be free to develop its economy with significantly less dependence on others. The costs of food production would fall rapidly. Suddenly, we would have more money for health and education. Presumably there would still be a few corrupt dimwits in government who would seize on the idea of invading other countries. But they would have less and less justification for their aggression. The world could also stop fighting over water because seawater could be desalinated so cheaply that it wouldn't be worth charging for drinking water. No one flees in rubber rafts from stable economies with a low carbon footprint and decent education and health systems. Even the unstoppable effects of climate change would weigh less heavily. Nearly free energy could cool and heat living space as needed and increase individual climate resilience all over the world, as can already be seen today in the model eco-cities of the Arabian Peninsula.

In terms of technology, abundant CO_2-free energy isn't a pipe dream. It could be achieved in many countries in ten years and

IF WE SOLVE
THE ENERGY
PROBLEM,
A WHOLE
HOST OF
ADDITIONAL
GLOBAL
PROBLEMS
WILL ALSO
VANISH.

become a global reality in twenty years at the latest. In many locations around the world, solar and wind energy can already be produced so cheaply that gas-, oil-, and coal-fired power plants don't make sense economically. The news has gradually gotten around, but people still often overlook how the price decline is continuing uninterrupted with a speed that we've seldom seen in the history of technology. Solar panels are becoming less and less expensive, while offshore wind farms are becoming less costly and more reliable. A SPRIND project is developing a 300-meter (1,000-foot) tall onshore windmill using a new type of tubular steel construction for the constant winds at altitude and a generator on a rotating base plate.[2]

We're gradually approaching a production price of 2 eurocents per kilowatt-hour of electricity for renewable energies. The most efficient solar power plants in Saudi Arabia and Abu Dhabi can do this trick for even 1 US cent.[3] Is it still worth billing customers for it? At the very least, it's worthwhile to convert electricity into hydrogen, natural gas, biofuels, or some other climate-neutral form during times of peak use and when consumption is low. In the technical jargon of environmental engineers, this is referred to as power-to-x. Power-to-x will provide energy for the sectors of a post–fossil fuel transportation system that are difficult to electrify. Travel above Earth's atmosphere will be one of them.

We're currently seeing considerable creativity and progress in practically every field of storage technology. Some ideas seem like they were hatched by a ten-year-old with advanced Lego skills. Huge concrete blocks suspended from cranes on steel cables are pulled upward using wind power. When the wind is still, the blocks release their energy again on the way down. You could hollow out mountains to create a water storage cylinder inside them, keep solar

WE KNOW HOW FUSION WORKS AND HOW WE CAN USE IT. WHETHER WE FULLY DEVELOP IT IS A MATTER OF HOW MUCH EFFORT WE CHOOSE TO INVEST.

power available at night by using the kinetic energy of large flywheels, or store energy using compressed air in salt domes that are no longer in use. Using the salt dome plan, wind energy would—literally—be available at any time. These ideas may sound crazy precisely because physically they seem so simple. But maybe we shouldn't dismiss them too quickly, even if battery development is progressing quite well at the moment. Why is it making progress now? Because for the last fifteen years, it's been an area of intensive research again.

Public attention is often focused on high-end batteries that achieve high energy densities by using expensive metals. However, we're especially interested in the rising learning curve and the fall in prices for heavy, inexpensive, and more environmentally friendly battery technologies, such as sodium-sulfur technology for stationary use. And even if it's inspiring that John Goodenough, one of the inventors of the lithium-ion battery, is still conducting research and applying for patents at the age of almost 100, in all likelihood, there will still be a residual demand for base load electricity in 2050 that we won't be able to meet using storage facilities.

Nuclear energy, whether some want it or not, will very likely also play an important role in a CO_2-neutral future. The next step in development will be small, extremely safe reactors that are quicker to build, thanks to modular design. Among others, Rolls-Royce is expected to bring these "small modular reactors" online in the UK in the early 2030s. Alternative designs, such as Natrium reactors, could also become commercial reality by the end of the decade, if Bill Gates's company TerraPower succeeds with its demonstration nuclear power station currently built in the US state of Wyoming.[4] Nuclear fusion based on new reactor technologies remains the even greater hope. We know how fusion works and how we can use it. Whether we fully develop it is a matter of the effort we choose to invest. The current acceleration of development is driven strongly by start-ups and entrepreneurs and not only by grand government-funded projects, a strong signal that finally nuclear fusion will not remain a moving target, which will always be ready in thirty years from any given moment. Incidentally, the same applies to efforts to recycle or neutralize nuclear waste, whether through so-called transmutation processes (by bombarding the highly radioactive residues with neutrons in next-generation fast breeder reactors) or, even more promising, through new laser-based processes, such as those currently being tested by the physics Nobel Prize winners Gérard Mourou and Donna Strickland.

THE SECOND CAMBRIAN EXPLOSION

How did our parents and grandparents imagine the future? It's striking how many retro-futuristic illustrations from the 1950s depict the future of transportation and mobility: flying cars and small, self-driving capsules in traffic-free zones, rocket cars and supersonic passenger planes, aerial trams between urban skyscrapers, and even intercontinental pneumatic tube systems for people. The capsules are highly reminiscent of Google's egg-shaped self-driving cars, the Concorde is no longer flying, and we suspect that traveling in Hyperloop will remain a technological pipe dream at least for our lifetimes because the physics of the human-sized pneumatic tube just don't work out. But above all, a look at the history of the future suggests that our generation too is probably extrapolating its technological future too much as a continuation of the past.

Technological development in our lifetime has been shaped most strikingly by the development of information and communication technology. In the visions of the future discussed today, artificial intelligence usually plays the leading role. The more hopeful variant of this vision highlights the all-knowing and ever-helpful computer assistant for our everyday lives. Many nerds see a potential technological utopia in the form of cyborgs with chips implanted in their brains to connect human intelligence directly to artificial intelligence. Even in Silicon Valley, the dystopian vision of an artificial general intelligence that frees itself from human control and then eradicates us, subjugates us, or keeps us in zoos for entertainment is seen as unlikely. Nobody can know what computers will be capable of a century or two from now, but we do know that no current path of technological development seems capable of creating either a benevolent or a malicious artificial intelligence. When it comes to the future advances or development possibilities of communications technology, at least of the directly visible or perceivable type, we're probably currently committing an error of linear overestimation similar to how the futuristic illustrations of the 1950s treated transportation technology. In the meantime, the growing wealth of data and computational capacity in chemistry and biotechnology could contribute to the kind of spurts in development that our generation has previously experienced with microelectronics and computers and their impact on work and value creation. As field-spanning innovations, new computers will make this "Cambrian explosion" possible. What lies in store for us?

When we say "computer" today, we have in mind a digital computer architecture like the one John von Neumann came up with seventy-five years ago. Driven by Moore's law, their electronic circuits have been growing smaller and increasingly powerful for more than fifty years. At the moment, 100 million transistors fit on a square millimeter, or over 5 million in a space a hair's breadth in diameter. Although it's been repeatedly declared dead, the vital signs of Moore's law are as strong as ever.

However, digital computers of von Neumann's type have some major drawbacks. These include susceptibility to software errors and thus to cyberattacks, as well as high power consumption and the associated heat dissipation. In the von Neumann architecture, the algorithms also work through their tasks "step by step." In many applications, it would be better to calculate things in a different way, especially for the simulation of natural processes, such as brain activity, molecular reactions, and fluid dynamics. In the transition from the digital to the real, analog world, the strengths of natural analog computers come into play. This can be illustrated by a not entirely serious example devised by Richard E. Borcherds involving a porcelain teapot.[5] Even the most powerful digital supercomputers can't precisely simulate what will happen when the teapot falls from someone's hand and shatters on a tile floor. It can only be approximated, an exercise that requires an enormous expenditure of energy. The solution to this computational dilemma would be a "teapot computer," that is, a teapot that we drop on the floor. Nature performs the calculations at lightning speed and with 100 percent precision and ultralow energy consumption each time (not including the energy required for cleaning the floor). The teapot computer isn't very practical—it can only do one trick, and after that it's irreparably broken. But it does illustrate the principle behind analog computers quite well.

Analog computers are specialists. For example, if developers need to calculate fluid dynamics, they build an electronic circuit that implements the formula for calculating flow in an analog electronic circuit. They let the electrons flow inside it, and they continuously measure the result. Every electron contributes to the result. Not one electron is wasted. This makes analog computers as energy efficient as our brains. Despite their estimated 100

WE ARE PROBABLY EXTRAPOLATING OUR TECHNOLOGICAL FUTURE TOO MUCH AS A CONTINUATION OF THE PAST.

billion neurons and 100 trillion connections, our brains consume just 20 watts of electricity and thus could be run on your phone charger.

Quantum computers are a special form of analog computer, with the difference being that the properties of individual physical particles that are difficult for humans to understand are used for calculation: the indeterminacy of electron states and entanglement between particles. With quantum computers, we're letting a single electron "dance." If that works, quantum computing will be used where traditional supercomputers fail. They will search extremely large amounts of data more quickly and more accurately, improve cryptography (with the risk that current encryption methods will no longer be secure), and optimize complex systems, for example, in logistics. As far as applications go, it's not especially important which types of analog computers prevail in the coming years. It seems certain that wherever biology, chemistry, or physics gets too complicated or intricate, analog computers will achieve great things. In simulations and calculations dealing with atoms and molecules and at the intersection of chemistry, biology, and physics, they will generate new insights into the reactive behavior of materials and cells, which will help materials science and the life sciences make new leaps in progress. It is highly likely that what we've seen with so many other advances in information technology over the past two decades will be repeated in this area as well.

With machine learning, we will be better able to predict how proteins fold. This will create entirely new possibilities for the development of medical compounds. Chemistry informatics is already revolutionizing the calculation of catalysis in some areas. This not only helps us in battery technology and in the search for ultralight, ultrastrong materials that are also 100 percent recyclable. We are also already seeing tremendous breakthroughs with bioreactors, such as in the production of lipids for mRNA vaccines. Research is being carried out into processes for converting rapidly growing plants into building materials for high-rise buildings. This would provide an inexpensive way to extract CO_2 from the air and store it permanently in buildings. It would also reduce the use of concrete, although concrete can be made significantly more environmentally friendly if it's strengthened with carbon fiber instead of steel. We suspect that it will be easier to extract carbon dioxide from the air with plants than with giant carbon dioxide capture facilities, but we would be happy to be proved wrong, especially if the capture facilities not only absorb CO_2 but also remove other harmful particles from the air.

Fast-growing plants that human beings can use as a tool to fight climate change probably don't even require genetic engineering. Evolution has already produced a number of interesting species, for example the empress tree, which rapidly deposits carbon from the air in its wood.[6] Nevertheless, we still wish for less ideological baggage when considering the potential of genetically modified plants, particularly in combination with high-tech agriculture, perhaps even vertical farming, to radically reduce the amount of land used for food production by 2050 and to create space for biodiversity and ecofriendly housing. The same of course applies to artificial meat grown in giant petri dishes or extruded from a bioprinter. Even organic meat is terrible for the climate, after all.

The use of genetic engineering and biotechnology outside the plant kingdom could lead to tremendous advances as well as the most profound ethical conflicts and risks. With the toolkit offered by CRISPR and other genetic engineering tools, we can responsibly carry out genetic interventions where nature now inflicts suffering or death on human life. With any luck, over the next few decades we'll experience a science-led, humanmade Cambrian explosion thanks to models and empirical studies in all the life sciences—in microbiology, in medicine, and in any field where we're coming to a better understanding of what holds our world together on a molecular level. The platforms for major breakthroughs using the transmission electron microscope have already been built, or construction is at least underway. Mass data analysis in genetic databases will help us finally understand the major diseases, and also the rare ones. We will produce many substances in batches of one dose and finally have precision medicine worthy of the name: the right compound for each individual and each pathology, individually printed—often in a decentralized fashion at the pharmacy around the corner.

This doesn't mean that we won't still be fighting disease in 2050. Perhaps we'll have to fight new diseases we haven't discovered yet. This isn't just a problem of the "unknown unknowns" and "black swans" that we're familiar with from research on possible future scenarios.[7] Unfortunately, nature has apparently integrated entropy, the creative-destructive process of decay, deeply into its structural design. But we will remain healthier as we grow old. Some biblical promises have a good chance of being fulfilled, at least the medical ones. Many deaf people will be able to hear in the future, and many blind people will be able to see again. Computer–brain interfaces will help paralyzed people walk again. We will be able to cure many types of cancer and autoimmune diseases, contain dementia, and most likely defeat Parkinson's disease. We will probably stay young longer by significantly slowing down the cellular aging process, perhaps even turning it off completely.

In 2050, the authors of this book, if we're still alive, will be over eighty years old. If, despite advances in the life sciences, we're dependent on nursing care, it will be to a significant degree digital and robotized. We see this as an appealing vision of the future, since we'd rather be washed by a robot than by a stressed-out caregiver. Nurses have better things to do and will hopefully be paid better for tasks requiring empathy than they are today for physically demanding tasks. Of course, it would be even nicer if we could go to the zoo in perfect health with our grandchildren and look at a woolly mammoth or a dodo bird.

SOME BIBLICAL PROMISES HAVE A GOOD CHANCE OF BEING FUL-FILLED, AT LEAST THE MEDICAL ONES.

Jurassic Park is not just a fantasy based on the vaguely scientific schemes of Hollywood writers.[8] It is approaching the realm of possibility that researchers will be able to resurrect extinct animal species if their DNA and some exogenetic material are still present.

DISRUPTING EDUCATION

But for our grandchildren, we wish for one thing most of all: for their schools to systematically promote their curiosity and intrinsic motivation to learn. Typically, discussions of breakthrough innovation ignore developments in pedagogy, teaching, and curriculum in our school systems. However, our impression is that we need an education disruption. Teaching and learning have to improve exponentially and become available to all. Only then will we as a society foster enough creative minds who will be able to think up new innovative leaps and develop them as technology.

What is stopping us from doing that? Discussions of educational problems over the last few years (and decades) have focused on incremental improvements. How do we improve teacher training? Should schools devote more class time to STEM topics, and should computer programming be a required course? Should the priority be schools for all ability levels and career plans, or should we focus on preparing students for college? The arguments have been endless, as has been the time spent in working groups, with little to show for it. As we see it, however, the elephant in the room has hardly been mentioned. Children come to school motivated. They're curious. For them, learning is fun. But by the time they hit eighth to tenth grade, many of them have completely lost interest. Their curiosity is gone.[9] Extrinsic incentives help the ones with self-discipline keep at it to some extent. Those who don't have self-discipline (or whose parents don't provide it) muddle their way through—and that's the best case. In the worst case, they fall by the wayside. The vastly declining levels of fun and motivation can't be fully explained by hormone levels during adolescence. After all, that's something every generation has experienced. Teaching and curriculum are out of touch and out of date. We'll only be able to get the elephant *out* of the room if the education system discards expensive, boring, and outdated textbooks; learns from the best game designers and visual artists; imparts explicit knowledge like the best YouTubers; implements adaptive learning software; and otherwise does what it tells students to do: pursue a path of continuous self-development.

What could the school of the future look like? The most fundamental change would consist of digitizing all subjects that focus on the traditional transfer of knowledge. This would free up time, money, and space for significantly smaller learning groups for all forms and formats of peer and discursive learning in all relevant subjects. Specifically, that means the best teachers would make the best videos that take students along on exciting learning journeys like a good Netflix series. Perhaps the best video teachers aren't trained teachers at all, but rather practitioners, actors, or students themselves. In digital systems, you can experiment and measure the results. Competition stimulates the digital learning environment. The new generation of video teachers will be able to draw on the best infographics, animations, and visual effects. That will be expensive, but in contrast to a textbook that costs 50 dollars or (much) more, digital systems are scalable at almost no cost. The more students who use them, the less expensive they get. The videos and interactive exercises will build on each other like good adventure games, foster students' ambition to reach the next level of knowledge, and won't slow down students who want to devour a year's worth of material in a month. If a thirteen-year-old wants to, she will be able to work her way through the twelfth-grade curriculum in the evenings. And they will be affordable for everyone, probably free of charge.

Exercise-based learning will also be much more centralized in hybrid teaching models combining AI assistance and online coaching by teachers. Adaptive learning software will customize itself to the knowledge and pace of learning of each individual student and serve up math and English assignments at the appropriate level. Software designers will master the tricks that game designers use to motivate players to keep playing, because they will ensure that players will perceive how their learning is progressing. And if a student and his AI assistant get stuck, he can ping a teacher in the chat room and get things explained again via headset using a split screen.

Video tutorials and gamified learning software will unburden the system so that teachers can be fully used where they're really needed: as teachers and moderators of small groups in which a maximum of sixteen students really have the space to participate—and not in lectures to thirty-two students where anyone who didn't develop an extroverted personality in kindergarten gets lost in the crowd. Incidentally, a hybrid school doesn't have to start at 7:45 a.m., when the students with natural night owl biorhythms are barely awake. Students who aren't falling asleep would also likely increase teachers' motivation, although teachers would need to be selected with more attention to their intrinsic motivation. Perhaps in 2050 teachers will no longer enjoy lifetime tenure so that motivation-killing teachers can be systematically removed from the system.

It's clear to us that education systems are as slow to change as health systems. But we owe our children and grandchildren an innovative leap in this area. In a school with a lot more "Minecraft," we won't need to worry that screen time will rot kids' brains. The status

AN INNOVATIVE LEAP BEGINS WITH AN IDEA THAT OTHERS CONSIDER NONSENSICAL OR IMPOSSIBLE.

quo is destroying educational opportunities one after another because it ignores opportunities for digital learning one after another.

When we were growing up, conservative educators always said that television would make us stupid. Even back then, that was nonsense. In the 1970s and 1980s, students who watched documentaries, Hitchcock films, and the news could learn quite a lot. The digital learning opportunities of today and tomorrow will make the sorting effect of media even more pronounced. Curious and engaged children will find their magic window to knowledge and creativity on YouTube. Others will watch people blast aliens on Twitch. The most important responsibility of public education is to encourage curiosity about learning in all children. To make things better than they are now isn't asking for much. To make things significantly better, we need open innovation in a system that is now far too closed. And if we succeed in reinventing education, we can rest assured that the innovators of tomorrow will leap.

DON'T BE CAREFUL WHAT YOU WISH FOR

An innovative leap almost always begins with an idea that others consider nonsensical or impossible. So we need more nonsensical ideas. How do we find them? At the Federal Agency for Disruptive Innovation, Germany's ARPA, it has become a habit to ask applicants, potential cooperation partners, and podcast guests what innovative leap they would most like to see by the year 2050—and why they want to see that innovation in particular.

Here are ten of our favorites:

→ 1. Many people put the energy revolution high on their lists or at the very top. It's not immediately obvious that "too cheap to meter" energy addresses a lot of problems at their root. But it is obvious that we're swimming in an ocean of energy and are just too stupid to harvest and store it.

→ 2. A holodeck that can take us anywhere could dramatically reduce boring time-wasters like business travel. Could it one day replace vacation trips as well? Could we fly off to Hawaii on antigravity-powered aircraft, or just have ourselves beamed there directly? From today's perspective, the latter seems unlikely, but the preceding options don't.

→ 3. The Federal Agency for Disruptive Innovation is working on a drug to treat Alzheimer's disease. We need breakthroughs for all of humanity's scourges: for all types of cancer and cardiovascular diseases, depression and other serious mental illnesses, and also for obesity, hair loss (one of us has a personal stake in this), menstrual pain, and the common cold. Would all of these maladies please just go away?

→ 4. The urgently needed disruption of education will come. Steve Jobs once called the personal computer a "bicycle for the brain." What if we could connect computers directly to the brain? Would this give us the "kung fu download" from *The Matrix*, or a way to funnel information directly into our heads? While we're waiting for that, we'd take lightweight glasses that augment our real world with virtual information. Preferably today rather than tomorrow, we'd like to see fun

LET'S
GOOD

ANCE

BE

STORS

knowledge transfer games that excite children and reconnect adults to learning. Lecture-based frontal instruction can finally be a thing of the past.

5. The way Elon Musk keeps promising they'll be here next year, self-driving cars are threatened to become the new running gag in the history of technology. But why *cars*? The world has three dimensions, and there's much more room for maneuvering in the third dimension. Where are the completely autonomous, quiet, environmentally friendly flying vehicles that will free us from the space-wasting tyranny of the automobile?

6. How about a life in harmony with nature, but without giving up the things we like? Cheap energy is part of this, as are the holodeck and autonomous flying vehicles, because then we could declutter city centers and dispense with streets. We could produce meat and fish indistinguishable from nature and foodstuffs like palm oil, rice, wheat, corn, and others in small areas so we wouldn't need so much space for agriculture and could leave the seas in peace.

7. Wars becoming much more unlikely due to the positive effects on the lives of everyone thanks to the progress outlined here. Once we have less war we can spend much less money on military products, making the resources available for more useful products. A democratic electoral system that attracts the most capable people into government and legislative service. A law enshrined in the Constitution for the legal code's maximum total word count. Perhaps every new law would have to abolish two old ones. An automatic expiration date for laws. Anything not actively extended would expire. Govern-

ment administrations oriented to results rather than process, unlike the one we have now (its motto: the surgery was conducted by the book; the patient, however, is dead).

→ 8. Social media that doesn't reinforce the worst qualities in people, but rather the best ones. Unfortunately, our long-cherished hope that the trolls would all kill each other off has not been fulfilled. We need generally recognized, broadly used digital platforms on which democratic discourse can be conducted in an orderly and structured manner. Arguments can be exchanged and further developed on them in depth and with the necessary chronological distance. Discussions on the platform would be moderated by wise, polite people with balanced personalities. Facts would be facts here, not fakes. Misplaced journalistic focus on the wrong balance between opposing viewpoints rather than on evidence would be abolished. Candidates in the next general election would either participate on the platform—or they would have no chance of being elected.

→ 9. A huge sun blind at Lagrange point 1, with which we can dim the solar radiation on Earth so finely granulated that we lower the temperature without triggering chaotic side effects in climate and weather. Of course, we too are aware of the dilemmas associated with solar geoengineering and how careful we have to be as long we don't understand climate models sufficiently. But it seems utterly negligent to us not to explore the technological possibilities with full scientific energy in an international collaboration and regulation framework. If we don't, there is a high probability that we will see individual countries experiment on their own in geoengineering—with unforeseeable risks for all of humanity.

→ 10. A colony on Mars with at least 1,000 inhabitants. What's the point of sending people to Mars without a return ticket, leaving them to live there in sealed environments, maybe even underground? Perhaps a Mars colony would restore to humanity the spirit of discovery that Marco Polo and Vasco da Gama had. It's what we'll need if we're going to be able to leave our solar system at some point, headed for the exoplanets if that becomes necessary to ensure the survival of humanity. Wildcard: Contact with or a visit from aliens. Nothing could help us progress like they could.

TECHNO-OPTIMISM AND SELF-EFFICACY

What does our technological future look like? The advances in information technology over the past few decades have brought about an interesting paradox. Thanks to massive quantities of data, refined mathematical models, and machine learning, we can predict linear developments much better than the generations before us. Based on abundant data, we've built a whole series of "prediction machines," as the economists Ajay Agrawal, Joshua Gans, and Avi Goldfarb call Big Data analytics and learning algorithms.[10] But the leaps in innovation largely induced by digital technologies are currently hatching all kinds of black swans. In futurology, black swans are phenomena and developments that nobody expects because they were unknown in the past. The consequence of this is that we also can't predict our future. It remains literally unpredictable and isn't technologically determined. That's a good thing, because this means we can use technology to create a positive version of the future. We can use technology to think of appealing future scenarios, then do everything we can to make as many positive elements of our vision as possible a reality as the future becomes the present. Perhaps Abraham Lincoln said it first, but certainly the computer scientist Alan Kay got right to the heart of the issue: "The best way to predict the future is to invent it."[11]

Forecasts can help. But what counts is the will to create. The purpose and objective of innovation is the greatest possible happiness for the greatest possible number of people. Because we believe in the will and ability of human creation, we are rationally optimistic that in the year 2050, with radically better technology, humanity will have many of the problems under control that today seem almost impossible to solve and sometimes even seem to threaten our continued existence. Pessimism, with or without technology, is essentially a waste of mental energy. Maybe here or there you can find positive side effects of the world-is-ending attitude. Managers in the tech industry like to quote *Only the Paranoid Survive*, the book by Intel cofounder Andy Grove.[12] But apocalyptic attitudes seem to have reached their point of marginal utility a few years ago. Anxiety disorders will not show us the path to a successful future. They lead instead to status quo bias and unhealthy risk aversion. Fear of the future is also at the root of excessive regulation.

Another concept from psychology seems much more helpful to us. We need to develop more collective self-efficacy in technological development, because self-efficacy has self-reinforcing effects. We achieve innovative leaps because we believe in ourselves. Because we believe in ourselves, we achieve innovative leaps. An optimistic vision of the future and a positive self-image go hand in hand with a bright future. This will also help us reduce excessive regulation and red tape.

Techno-optimism is not an end in itself, but an ethical imperative. The goal is not as difficult to define as we often think. Jonas Salk, the American physician, immunologist, and developer of the inactivated polio vaccine, described the goal as follows: "Our greatest responsibility is to be good ancestors."[13]

Our descendants should have
an even better life than we do.
To accomplish this, we have
to be optimistic and set out for
a leap over the brink of utopia
to make the world greener,
healthier, and wealthier.

NOTES

Prelude

[1] Ernst Bloch, *The Principle of Hope* (Cambridge, MA: MIT Press, 1986).

Chapter 1

[1] Yuval N. Harari, *Sapiens: A Brief History of Humankind* (London: Harvill Secker, 2014).

[2] James Fallows, "The 50 Greatest Breakthroughs since the Wheel," *The Atlantic*, November 15, 2013, https://www.theatlantic.com/magazine/archive/2013/11/innovations-list/309536/.

[3] Tyler Cowen, *The Great Stagnation: How America Ate All the Low-Hanging Fruit of Modern History, Got Sick, and Will (Eventually) Feel Better* (New York: Dutton, 2011); Robert Gordon, The *Rise and Fall of American Growth: The U.S. Standard of Living since the Civil War* (Princeton, NJ: Princeton University Press, 2016); Lee Vinsel and Andrew Russell, *The Innovation Delusion: How Our Obsession with the New Has Disrupted the Work That Matters Most* (New York: Crown Publishing, 2020).

[4] Robert Hone and John Palfreman, "Interview with Steve Jobs, 1990," GHB Archives, May 14, 1990, video, 02:40–03:12, https://openvault.wgbh.org/catalog/V_AD9E0B-C353BF435E83F28DEF165D4F40.

[5] Ryan Avent, "Everything's Amazing and Nobody's Happy," *The Economist*, September 11, 2012, https://www.economist.com/free-exchange/2012/09/11/everythings-amazing-and-nobodys-happy.

[6] Vinsel and Russell, *The Innovation Delusion*, 37.

[7] Kari Paul, "Elizabeth Holmes Trial: Jury Finds Theranos Founder Guilty on Four Fraud Counts," *The Guardian*, January 4, 2022, https://www.theguardian.com/technology/2022/jan/03/elizabeth-holmes-trial-jury-finds-theranos-founder-guilty-on-four-counts.

[8] Nicholas Bloom et al., "Are Ideas Getting Harder to Find?" *American Economic Review* 110, no. 4 (September 2017): 1104–1144.

[9] Thomas Ramge, "Wie nützt das Neue?" *brand eins*, July 2013, https://www.brandeins.de/magazine/brand-eins-wirtschaftsmagazin/2013/fortschritt-wagen/wie-nuetzt-das-neue.

[10] Elizabeth Pain, "French Science Bill Promises Boost to Public R&D," *Science*, July 24, 2020, https://www.science.org/content/article/french-science-bill-promises-boost-public-rd.

[11] "20 Milliarden für Bildung und Forschung," Bundesministerium für Forschung und Bildung, June 2, 2022, https://www.bmbf.de/bmbf/shareddocs/kurzmeldungen/de/2022/05/bmbf-haushalt-2022.html.

12 "Horizon Europe," European Commission, accessed September 17, 2022, https://research-and-innovation.ec.europa.eu/funding/funding-opportunities/funding-programmes-and-open-calls/horizon-europe_en.

13 The White House, "The Biden-Harris Administration FY 2023 Budget Makes Historic Investments in Science and Technology," White House Press Releases, April 5, 2022, https://www.whitehouse.gov/ostp/news-updates/2022/04/05/the-biden-harris-administration-fy-2023-budget-makes-historic-investments-in-science-and-technology/.

14 McKenzie Prillaman, "Billions More for US Science: How the Landmark Spending Plan Will Boost Research," *Nature* 608, no. 7922 (August 2022): 249, https://www.nature.com/articles/d41586-022-02086-z; Rebecca Leber, "The US Finally Has a Law to Tackle Climate Change," Vox, last modified August 16, 2022, https://www.vox.com/policy-and-politics/2022/7/28/23281757/whats-in-climate-bill-inflation-reduction-act.

15 PRC State Council, "China's Spending on R&D Hits 3 Trln Yuan in 2022," The People's Republic of China (PRC) State Council Statistics, last modified January 23, 2023, http://english.www.gov.cn/archive/statistics/202301/23/content_WS63ce3d-b8c6d0a757729e5fe5.html

16 UNESCO Institute for Statistics, "Research and Development Expenditure (% of GDP)," World Bank, accessed September 18, 2022, https://data.worldbank.org/indicator/GB.XPD.RSDV.GD.ZS.

17 Lewis M. Branscomb, *Investing in Innovation: Creating a Research and Innovation Policy That Works* (Cambridge, MA: MIT Press, 1998).

18 EU Joint Research Centre, "China Overtakes the EU in High-Impact Publications," The European Commission's science and knowledge service, March 3, 2021, https://joint-research-centre.ec.europa.eu/system/files/2021-05/policy_brief_china_overtakes_the_eu_in_high_impact_publications_final_jrc124597.pdf.

19 Viktor Mayer-Schönberger and Thomas Ramge, *Access Rules: Freeing Data from Big Tech for a Better Future* (Berkeley: University of California Press, 2022).

20 Peter Thiel and Blake Masters, *Zero to One* (New York: Random House, 2014).

21 Shirin Ghaffary, "Big Tech Has a Battle Ahead over Antitrust Regulation—and It's Going to Get Messy," *Vox*, June 18, 2021, https://www.vox.com/recode/22537529/tech-battle-antitrust-regulation-lina-khan-ftc-google-facebook-apple-amazon-cicilline-congress; Adi Robertson, "How the EU Is Fighting Tech Giants with Margrethe

Vestager," *The Verge*, March 17, 2022, https://www.theverge.com/22981261/margrethe-vestager-decoder-antitrust-eu-apple-facebook-google-jedi-blue.

22 Jeff Tollefson, "The Rise of 'ARPA-Everything' and What It Means for Science," *Nature* 595, no. 7868 (July 2021): 483–484.

23 Brian Owens, "Canada Announces New Innovation Agency—and It's Not Modelled on DARPA," *Nature*, April 27, 2022, https://www.nature.com/articles/d41586-022-01190-4.

24 Tollefson, "The Rise of 'ARPA-Everything,'" 483–484.

25 Nikolai D. Kondratieff, "The Long Waves in Economic Life," *The Review of Economics and Statistics* 17, no. 6 (November 1935): 105–115.

26 Joseph A. Schumpeter, *Business Cycles* (New York: McGraw-Hill, 1939).

27 Freeman Christopher, John Clark, and Luc Soete, *Unemployment and Technical Innovation: A Study of Long Waves and Economic Development* (Westport, CT: Greenwood Press, 1982).

28 Mark Weiser, "The Computer for the 21st Century," *Scientific American* 265, no. 3 (1991): 94–104.

29 "Top 250—Global Energy Company Rankings," S&P Global, accessed September 20, 2022, https://www.spglobal.com/commodityinsights/top250/rankings.

30 Liu Cixin, *The Three-Body Problem* (New York: Tor Books, 2014).

Chapter 2

1 Mar Fernández, Interview by Thomas Ramge, *SPRIND Podcast*, Federal Agency for Disruptive Innovation SPRIND, July 4, 2022, audio, 33:39. https://www.sprind.org/de/podcast/36-mar-fernandez/.

2 Jill Lepore, "The Disruption Machine: What the Gospel of Innovation Gets Wrong," *New Yorker*, June 16, 2014, https://www.newyorker.com/magazine/2014/06/23/the-disruption-machine.

3 Jeremy Bentham, *An Introduction to the Principles of Morals and Legislation* (London: T. Payne and Sons, 1780).

4 Thiel and Masters, *Zero to One*.

5 "The Foie Gras'ing of Startups: Does Raising More VC Lead to Bigger Outcomes?" CBInsights, July 10, 2019, https://www.cbinsights.com/research/startup-vc-overfunding-foie-gras/.

6 Lizzie Widdicombe, "The Rise and Fall of WeWork," *New Yorker*, November 6, 2019, https://www.newyorker.com/culture/culture-desk/the-rise-and-fall-of-wework.

7 "Transforming Our World: The 2030 Agenda for Sustainable Development," United Nations, accessed September 18, 2022, https://sdgs.un.org/2030agenda.

[8] Abraham H. Maslow, Henry Geiger, and Bertha G. Maslow, *The Farther Reaches of Human Nature* (New York: Compass, 1971).

[9] Philine Warnke et al., "100 Radical Innovation Breakthroughs for the Future: The Radical Innovation Breakthrough Inquirer," European Commission, June 24, 2019, https://euraxess.ec.europa .eu/worldwide/india/ec-publishes -forsesight-study-100-radical -innovation-breakthroughs-future.

[10] Matt Ridley, *How Innovation Works: And Why It Flourishes in Freedom* (New York: HarperCollins, 2020), 39.

[11] WHO, *Global Status Report on Road Safety 2018* (Geneva: World Health Organization, 2018), 3.

[12] David Banta, "What Is Technology Assessment?" *International Journal of Technology Assessment in Health Care* 25, no. S1 (2009): 7–9, https://doi.org/10.1017/ S0266462309090333.

[13] David Collingridge, *The Social Control of Technology* (London: Pinter, 1980).

[14] John R. Kimberly and Michael J. Evanisko, "Organizational Innovation: The Influence of Individual, Organizational, and Contextual Factors on Hospital Adoption of Technological and Administrative Innovations," *Academy of Management Journal* 24, no. 4 (December 1981): 689–713.

[15] "About Thalidomide," The Thalidomide Society, accessed September 21, 2022, https://thalidomide -society.org/what-is-thalidomide/.

Chapter 3

[1] "Framingham Heart Study," Boston University School of Medicine, accessed September 15, 2022, https://www.bumc.bu.edu/preventive -med/research/framingham-heart -study/.

[2] Safi Bahcall, *Loonshots: How to Nurture the Crazy Ideas That Win Wars, Cure Diseases, and Transform Industries* (New York: St. Martin's Press, 2019), 45–56; Akira Endo, "A Gift from Nature: The Birth of the Statins," *Nature Medicine* 14 (October 2008): 1050–1052; Akira Endo, "A Historical Perspective on the Discovery of Statins," *Proceedings of the Japan Academy. Series B, Physical and Biological Sciences* 86, no. 5 (2010): 484–493, https://doi.org/10.2183/ pjab.86.484.

[3] Malcolm Gladwell, *Outliers: The Story of Success* (New York: Little, Brown, 2008).

[4] Barbara Goldsmith, *Obsessive Genius: The Inner World of Marie Curie* (New York: W. W. Norton, 2005); Peter Theiner and Robert Bosch, *Unternehmer im Zeitalter der Extreme* (Munich: C. H. Beck, 2019); Walter Isaacson, *Steve Jobs* (New York: Simon & Schuster, 2011); Ashlee Vance, *Elon Musk: Tesla, SpaceX, and the Quest for*

a Fantastic Future (New York: HarperCollins, 2015).

5 Adam Grant, *Originals: How Non-conformists Move the World* (New York: Viking, 2016); Kevin Ashton, *How to Fly a Horse: The Secret History of Creation, Invention, and Discovery* (New York: Doubleday, 2015); Matt Ridley, *How Innovation Works*; Bahcall, *Loonshots*.

6 *SPRIND Podcast*: Thomas Ramge spricht mit Menschen, die Neues neu denken," Federal Agency for Disruptive Innovation SPRIND, accessed September 23, 2022, https://www.sprind.org/de/podcast/.

7 Vinnie Mirchandani, *The New Polymath: Profiles in Compound-Technology Innovations* (Hoboken, NJ: John Wiley & Sons, 2010).

8 Angela L. Duckworth, *Grit: The Power of Passion and Perseverance* (New York: Scribner, 2016).

9 Anders Ericsson and Robert Pool, *Peak: Secrets from the New Science of Expertise* (New York: HarperCollins, 2017).

10 Bernd Ulmann, Interview with Thomas Ramge, *SPRIND Podcast*, Federal Agency for Disruptive Innovation SPRIND, April 26, 2021, audio, 50:44, https://www.sprind.org/de/podcast/7-bernd-uhlmann/.

11 Thomas Carlyle, *On Heroes, Hero-Worship, & the Heroic in History* (London: James Fraser, 1841).

12 AnnaLee Saxenian, *Regional Advantage: Culture and Competition in Silicon Valley and Route* 128 (Cambridge, MA: Harvard University Press, 1994).

13 Regina Dugan and Kaigham J. Gabriel, "Changing the Business of Breakthroughs," *Issues in Science and Technology* 38, no. 4 (Summer 2022): 70–74. See also Dan Wattendorf, Interview with Thomas Ramge, SPRIND Podcast, Federal Agency for Disruptive Innovation SPRIND, June 7, 2021, audio, 45:18, https://www.sprind.org/de/podcast/10-dan-wattendorf/.

14 Paul Graham, "The Bus Ticket Theory of Genius," November 2019, http://www.paulgraham.com/genius.html.

15 Simon Bowers, "Jeff Bezos: Amazon.com's 'Dread Pirate' Founder," *The Guardian*, August 6, 2013, https://www.theguardian.com/technology/2013/aug/06/jeff-bezos-amazon-washington-post.

16 Lucía Del Carpio and Maria Guadalupe, "More Women in Tech? Evidence from a Field Experiment Addressing Social Identity," IZA Institute of Labor Economics Discussion Paper Series no. 11876 (October 2018), https://docs.iza.org/dp11876.pdf.

17 "Labor Force Statistics from the Current Population Survey," US Bureau of Labor Statistics, accessed September 16, 2022, https://www.bls.gov/cps/cpsaat11.htm.

18 National Science Foundation, "Science and Engineering Degrees: 1966–2010," National Center for

Science and Engineering Statistics, last modified June 2013, https://www.nsf.gov/statistics/nsf13327/content.cfm?pub_id=4266&id=2. See also "Women in Computer Science: Getting Involved in STEM," ComputerScience.org, last modified September 9, 2022, https://www.computerscience.org/resources/women-in-computer-science/.

[19] Steve Henn, "When Women Stopped Coding," *NPR Morning Edition*, NPR, October 21, 2014, audio, 4:33, https://www.npr.org/sections/money/2014/10/21/357629765/when-women-stopped-coding?t=1660301924425&t=1660302339887.

[20] Caroline C. Perez, *Invisible Women: Exposing Data Bias in a World Designed for Men* (New York: Vintage, 2020).

Chapter 4

[1] Ulrich Eberl, *Smarte Maschinen: Wie künstliche Intelligenz unser Leben verändert* (Munich: Hanser, 2016), 11–16.

[2] John Markoff, "Japanese Team Dominates Competition to Create Generation of Rescue Robots," *New York Times*, December 22, 2013, https://www.nytimes.com/2013/12/23/science/japanese-team-dominates-competition-to-create-rescue-robots.html.

[3] Gill Pratt, "DARPA Challenge History," Humanoid Robot, January 1, 2014, YouTube 02:40–03:03, https://www.youtube.com/watch?v=MGMk4SdLGjE.

[4] Madeline Verniero, "Does DARPA's Legal Structure Lead to Ethical Lapses?" *Regulatory Review*, June 16, 2022, https://www.theregreview.org/2022/06/16/verniero-does-darpas-legal-structure-lead-to-ethical-lapses/.

[5] William B. Bonvillian, Richard Van Atta, and Patrick Windham, eds., *The DARPA Model for Transformative Technologies* (Cambridge: Open Book Publishers, 2019).

[6] DARPA, "DARPA 2019 Strategic Framework," Defense Advanced Research Projects Agency, accessed September 19, 2022, https://www.darpa.mil/attachments/DARPA-2019-framework.pdf.

[7] William B. Bonvillian, "The Connected Science Model for Innovation—the DARPA Model," in *The DARPA Model for Transformative Technologies*, ed. William B. Bonvillian, Richard Van Atta, and Patrick Windham (Cambridge: Open Book Publishers, 2019), 77–144.

[8] William B. Bonvillian, "Power Play," *American Interest* 2, no. 2 (November 1, 2006), https://www.the-american-interest.com/2006/11/01/power-play/.

[9] Alex Davis, "The Autonomous-Car Chaos of the 2004 DARPA Grand Challenge," *Wired*, January 6, 2021, https://www.wired.com/story/autonomous-car-chaos-2004-darpa-grand-challenge/.

[10] Margaret O'Mara, *The Code: Silicon Valley and the Remaking of America* (New York: Penguin Press, 2019).

[11] Steven Overly, "This Government Loan Program Helped Tesla at a Critical Time: Trump Wants to Cut It," *Washington Post*, March 16, 2017, https://www.washingtonpost.com/news/innovations/wp/2017/03/16/this-government-loan-program-helped-tesla-at-a-critical-time-trump-wants-to-cut-it/. See also Chuck Squatriglia, "Feds Lend Tesla $465 Million to Build Electric Car," *Wired*, June 24, 2009, https://www.wired.com/2009/06/tesla-loan/.

[12] Mariana Mazzucato, *The Entrepreneurial State: Debunking Public vs. Private Sector Myths* (London: Penguin Allen Lane, 2013).

[13] Mazzucato, *The Entrepreneurial State*, 93–99.

[14] Mariana Mazzucato, *Mission Economy: A Moonshot Guide to Changing Capitalism* (London: Penguin Allen Lane, 2021).

[15] Max Weber, *Economy and Society* (Cambridge, MA: Harvard University Press, 2019); James Q. Wilson, *Bureaucracy: What Government Agencies Do and Why They Do It* (New York: Basic Books, 1991).

[16] "Positionspapier IP-Transfer 3.0," Federal Agency for Disruptive Innovation SPRIND, July 2022, https://www.sprind.org/cms/uploads/220423_IP_Paper_SPRIND_6f9b19a34f.pdf.

[17] "Consolidated Version of the Treaty on the Functioning of the European Union," *Official Journal of the European Union*, Section 2, Art. 107, No. 1, p. C326/91. (October 10, 2012), https://eur-lex.europa.eu/legal-content/EN/TXT/?uri=celex%3A12012E%2FTXT.

[18] Paul Zak, "Measurement Myopia," Drucker Institute, April 7, 2013, https://www.drucker.institute/thedx/measurement-myopia/.

[19] Graham Badley, "A Place from Where to Speak: The University and Academic Freedom," *British Journal of Educational Studies* 57, no. 2 (June 2009): 146–163.

[20] Donald E. Stokes, *Pasteur's Quadrant: Basic Science and Technological Innovation* (Washington, DC: Brookings Institution Press, 1997).

[21] Anna P. Goldstein and Venkatesh Narayanamurti, "Simultaneous Pursuit of Discovery and Invention in the US Department of Energy," *Research* 47, no. 8 (October 2018): 1505–1512, https://doi.org/10.1016/j.respol.2018.05.005.

[22] "Our Impact," Advanced Research Projects Agency–Energy (ARPA-E), accessed September 17, 2022, https://arpa-e.energy.gov/about/our-impact.

[23] Jeff Tollefson, "The Rise of 'ARPA-Everything,'" 483–484.

24 Daniel Kahneman and Amos Tversky, "Prospect Theory: An Analysis of Decision under Risk," *Econometrica* 47, no. 2 (1979): 263–291, https://doi.org/10.2307/1914185.

25 Ken Gabriel, "ZEIT für Forschung: Sparking Innovation through Challenges," ZEIT für Forschung, May 27, 2021, YouTube, 28:21–28:29, https://www.youtube.com/watch?v=ZJgrb1F1poE.

26 Daisuke Wakabayashi and Scott Shane, "Google Will Not Renew Pentagon Contract That Upset Employees," *New York Times*, June 1, 2018, https://www.nytimes.com/2018/06/01/technology/google-pentagon-project-maven.html.

27 Seth J. Frantzman, "Israel's Iron Dome Won't Last Forever," *Foreign Policy*, June 3, 2021, https://foreignpolicy.com/2021/06/03/israels-iron-dome-wont-last-forever/.

28 The OECD report on "Public Procurement for Innovation" offers a broad perspective on strategic approaches and a variety of best practices in many OECD countries: OECD, "Public Procurement for Innovation: Good Practices and Strategies," *OECD Public Governance Reviews*, June 2, 2017, https://www.oecd.org/gov/public-procurement-for-innovation-9789264265820-en.htm.

29 Amandine Le Pape, "$8.5M to Accelerate Matrix, Polish Riot and Expand Modular," *Element* (blog), October, 10, 2019, https://element.io/blog/8-5m-to-accelerate-matrix/; and "France Recruits Dassault Systemes, OVH for Alternative to U.S. Cloud Firms," Reuters, October 3, 2019, https://www.reuters.com/article/us-france-dataprotection-idUSKBN1WI189.

30 Liv McMahon, "Ministry of Defense Acquires Government's First Quantum Computer," BBC, June 9, 2022, https://www.bbc.com/news/technology-61647134.

31 "Bundesverwaltung beschafft maschinelles Übersetzungsprogramm," Swiss Federal Council, December 18, 2019, https://www.admin.ch/gov/de/start/dokumentation/medienmitteilungen.msg-id-77610.html.

32 Blake Matich and Jonathan Gifford, "Making Europe's Solar Future," *Green European Journal*, November 26, 2021, https://www.greeneuropeanjournal.eu/making-europes-solar-future/.

33 Jorge Luis García et al., "The Life-Cycle Benefits of an Influential Early Childhood Program," NBER Working Paper Series 22993, December 2016, http://www.nber.org/papers/w22993.

Chapter 5

1 Stefan Füssel, *Gutenberg* (London: Haus Publishing, 2019).

2 Jeff Jarvis, *Gutenberg the Geek* (Amazon, 2012), Kindle.

[3] Jon Cohen, "Gentlemen of Science," *Science* 279, no. 5348 (January 9, 1998): 179.

[4] Kat Eschner, "This Wooden Running Machine Was Your Fixie's Great-Great Grandpa," *Smithsonian Magazine*, February 17, 2017, https://www.smithsonianmag.com/smart-news/wooden-running-machine-was-your-fixies-great-great-grandpa-180962152/.

[5] Bundesgesetzblatt, "Gesetz für den Ausbau erneuerbarer Energien (Erneuerbare-Energien-Gesetz-EEG 2014)," Bundesgesetzblatt Jahrgang 2014 Teil I Nr. 33, July 21, 2014.

[6] Paul A. Gompers, "The Rise and Fall of Venture Capital," *Business and Economic History* 23, no. 2 (Winter 1994): 1–26.

[7] Tom Nicholas, *VC: An American History* (Cambridge, MA: Harvard University Press, 2019).

[8] David A. Kaplan, *The Silicon Boys and Their Valley of Dreams* (New York: William Morrow, 1999), 176.

[9] Rene Schäfer, "Das Kasino ist vorübergehend geschlossen," interview by Thomas Ramge, *brand eins*, July 2022, https://www.brandeins.de/magazine/brand-eins-wirtschafts-magazin/2022/wendepunkte/start-ups-in-der-krise-das-kasino-ist-voruebergehend-geschlossen.

[10] Massimo Portincaso et al., "The Deep Tech Investment Paradox: A Call to Redesign the Investor Model," Boston Consulting Group and Hello Tomorrow, May 2021, https://hello-tomorrow.org/wp-content/uploads/2021/05/Deep-Tech-Investment-Paradox-BCG.pdf.

[11] Esha Vaish and Simon Jessop, "Spotify Prompts Nordic Pension Funds to Add Private Equity to Playlists," Reuters, April 15, 2019, https://jp.reuters.com/article/us-nordics-privateequity-pensions-analys-idUSKCN1RR1QM.

[12] Paul Marks, "Where Have All the Start-Ups Gone?" *Technologist*, April 3, 2017, https://www.technologist.eu/where-have-all-the-start-ups-gone/.

[13] Margaret O'Mara, "Why Can't Tech Fix Its Gender Problem?" *MIT Technology Review*, August 11, 2022, https://www.technologyreview.com/2022/08/11/1056917/tech-fix-gender-problem/.

[14] Gené Teare, "VC Funding to Black-Founded Startups Slows Dramatically As Venture Investors Pull Back," Crunchbase, June 17, 2022, https://news.crunchbase.com/diversity/vc-funding-black-founded-startups/.

[15] O'Mara, "Why Can't Tech Fix Its Gender Problem?"

[16] Joseph A. Schumpeter, *Capitalism, Socialism and Democracy* (New York: HarperCollins, 1942).

[17] American Dynamism, "Our Investment Thesis," Andreessen Horowitz, accessed September 19, 2022,

https://a16z.com/american-dynamism/.

[18] Jane Edwards, "America's Frontier Fund to Invest in AI, Microelectronics to Secure US Tech Advantage," GovCon Wire, June 10, 2022, https://www.govconwire.com /2022/06/americas-frontier-fund-to -invest-in-ai-deep-tech-to-secure -us-tech-advantage/.

[19] Neal Stephenson, *Snow Crash* (New York: Bantam Books, 1992).

Chapter 6

[1] Leonard Kleinrock, "The First Internet Connection, with UCLA's Leonard Kleinrock," UCLA, January 13, 2022, YouTube, 02:55–04:17, https://www.youtube.com/watch?v =vuiBTJZfeo8&t=175s.

[2] Mark Weiser, "The Computer for the 21st Century," 94–104.

[3] Viktor Mayer-Schönberger and Thomas Ramge, *Reinventing Capitalism in the Age of Big Data* (New York: Basic Books, 2018).

[4] Marc Andreessen, "Why Software Is Eating the World," *Wall Street Journal*, August 20, 2011, https:// www.wsj.com/articles/SB10001424053 111903480904576512250915629460.

[5] "2020 Open Source Security and Risk Analysis Report," *Synopsys*, accessed September 19, 2022, https://ttpsc.com/wp3/wp-content/ uploads/2020/10/2020-ossra-report.pdf.

[6] "Public Money? Public Code!" Free Software Foundation Europe, accessed September 19, 2022, https://fsfe.org/activities /publiccode/publiccode.en.html.

[7] Microsoft, "Microsoft Acquires GitHub," Microsoft News Center, June 4, 2018, https://news.microsoft .com/2018/06/04/microsoft-to -acquire-github-for-7-5-billion/.

[8] Robert Loring Allen, *Opening Doors: The Life and Work of Joseph Schumpeter–Europe* (Piscataway, NJ: Transaction Publishers, 1991), 99.

[9] Viktor Mayer-Schönberger and Thomas Ramge, "The Data Boom Is Here–It's Just Not Evenly Distributed," *MIT Sloan Management Review*, February 9, 2022, https:// sloanreview.mit.edu/article/ the-data-boom-is-here-its-just-not -evenly-distributed/.

[10] "Data Act: Commission Proposes Measures for a Fair and Innovative Data Economy," European Commission, February 23, 2022, https:// ec.europa.eu/commission/presscorner/ detail/en/ip_22_1113.

[11] Mayer-Schönberger and Ramge, *Access Rules*, 91–93.

[12] Viktor Mayer-Schönberger and Thomas Ramge, "Are the Most Innovative Companies Just the Ones with the Most Data?" *Harvard Business Review*, February 7, 2018, https://hbr.org/2018/02/are-the -most-innovative-companies-just-the -ones-with-the-most-data.

[13] Peter Kurz, *Weltgeschichte des Erfindungsschutzes: Erfinder und Patente im Spiegel der Zeiten* (Cologne: Carl Heymanns, 2000).

[14] "America Wants to Waive Patent Protection for Vaccines," *The Economist*, May 8, 2021, https://www.economist.com/business/2021/05/08/america-wants-to-waive-patent-protection-for-vaccines.

[15] Ridley, *How Innovation Works*, 379.

[16] Ridley.

[17] "A Dispute over COVID-Vaccine Technology Ends Up in Court," *The Economist*, September 1, 2022, https://www.economist.com/business/2022/09/01/a-dispute-over-covid-vaccine-technology-ends-up-in-court.

[18] Alex Tabarrok, *Launching the Innovation Renaissance: A New Path to Bring Smart Ideas to Market Fast* (New York: TED, 2011).

[19] Henry W. Chesbrough, *Open Innovation: The New Imperative for Creating and Profiting from Technology* (Boston Harvard Business Publishing, 2003).

[20] Robert J. Allio, "CEO Interview: The InnoCentive Model of Open Innovation," *Strategy and Leadership* 34, no. 4 (August 2004): 4–9.

[21] Martin Watzinger et al., "How Antitrust Enforcement Can Spur Innovation: Bell Labs and the 1956 Consent Decree," *American Economic Journal* 12, no. 4 (November 2020): 328–359.

Chapter 7

[1] Steven Johnson, *How We Got to Now: Six Innovations That Made the Modern World* (New York: Riverhead Books, 2014).

[2] "Den Wind Ernten—Die Binnenwindanlagen der Zukunft," Federal Agency for Disruptive Innovation SPRIND, accessed September 19, 2022, https://www.sprind.org/de/projekte/beventum/.

[3] "Saudi PIF Launches Massive 1.5 GW Sudair Solar Energy Project," Arab News, April 9, 2021, https://www.arabnews.com/node/1840096/business-economy.

[4] "A Firm Founded by Bill Gates Bets On a Novel Nuclear Reactor," *The Economist*, June 10, 2021, https://www.economist.com/science-and-technology/2021/06/10/a-firm-founded-by-bill-gates-bets-on-a-novel-nuclear-reactor.

[5] Richard E. Borcherds, "The Teapot Test for Quantum Computers," February 7, 2021, YouTube, 02:53–10:50, https://www.youtube.com/watch?v=sFhhQRxWTIM.

[6] Emily Chasan, "We Already Have the World's Most Efficient Carbon Capture Technology," Bloomberg, August 2, 2019, https://

www.bloomberg.com/news/features
/2019-08-02/we-already-have-the
-world-s-most-efficient-carbon
-capture-technology?leadSource
=uverify%20wall.

[7] Nasim N. Taleb, *The Black Swan: The Impact of the Highly Improbable* (New York: Random House, 2007).

[8] Elizabeth Jones, "Sci-fi and Jurassic Park Have Driven Research, Scientists Say," The Conversation, June 10, 2015, https://theconversation.com/sci -fi-and-jurassic-park-have-driven -research-scientists-say-42864; Lorraine Boissoneault, "Jurassic Park's Unlikely Symbioses with Real-World Science," *Smithsonian Magazine*, June 15, 2018, https:// www.smithsonianmag.com/science -nature/jurassic-park-reveals-delicate -interplay-between-science-and -science-fiction-180969331/.

[9] Timo Gnambs and Barbara Hanf-stingl, "The Decline of Academic Motivation during Adolescence: An Accelerated Longitudinal Cohort Analysis on the Effect of Psychological Need Satisfaction," *Educational Psychology* 36, no. 9 (September 2016): 1698–1712.

[10] Ajay K. Agrawal, Joshua Gans, and Avi Goldfarb, *Prediction Machines: The Simple Economics of Artificial Intelligence* (Boston Harvard Business Publishing, 2018).

[11] NLM In Focus, "Dr. Alan Kay Talks About the Future at Annual Lindberg-King Lecture," National Library of Medicine, October 4, 2018, https://infocus.nlm.nih.gov /2018/10/04/dr-alan-kay-talks -about-the-future-at-annual -lindberg-king-lecture/. See also video at https://videocast.nih.gov/ Summary.asp?Live=28442&bhcp=1 ,quote at 4:13.

[12] Andrew S. Grove, *Only the Paranoid Survive: Lessons from the CEO of INTEL Corporation* (New York: Broadway Business, 1996).

[13] Bruce Alberts, "Science and Human Needs," 137th Annual Meeting, National Academy of Sciences, May 1, 2000, https://brucealberts.ucsf.edu/ publications/Speech7.pdf.

Advanced Research Projects Agency–Energy (ARPA-E). "Our Impact." Accessed September 17, 2022. https://arpa-e.energy.gov/about/our-impact.

Aghion, Philippe, Céline Antonin, and Simon Bunel. *The Power of Creative Destruction: Economic Upheaval and the Wealth of Nations*. Cambridge, MA: Harvard University Press, 2021.

Agrawal, Ajay K., Joshua Gans, and Avi Goldfarb. *Prediction Machines: The Simple Economics of Artificial Intelligence*. Boston, Harvard Business Publishing, 2018.

Alberts, Bruce. "Science and Human Needs." 137th Annual Meeting, National Academy of Sciences. May 1, 2000. https://brucealberts.ucsf.edu/publications/Speech7.pdf.

Allen, Robert Loring. *Opening Doors: The Life and Work of Joseph Schumpeter–Europe*. Piscataway, NJ: Transaction Publishers, 1991.

Allio, Robert J. "CEO Interview: The InnoCentive Model of Open Innovation." *Strategy and Leadership* 34, no. 4 (August 2004): 4–9.

American Dynamism. "Our Investment Thesis." Andreessen Horowitz. Accessed September 19, 2022. https://a16z.com/american-dynamism/.

"America Wants to Waive Patent Protection for Vaccines." *The Economist*, May 8, 2021. https://www.economist.com/business/2021/05/08/america-wants-to-waive-patent-protection-for-vaccines.

Andreessen, Marc. "Why Software Is Eating the World." *Wall Street Journal*, August 20, 2011. https://www.wsj.com/articles/SB10001424053111903480904576512250915629460.

Ante, Spencer E. *Creative Capital: Georges Doriot and the Birth of Venture Capital*. Cambridge, MA: Harvard Business Publishing, 2008.

Arab News. "Saudi PIF Launches Massive 1.5 GW Sudair Solar Energy Project." April 9, 2021. https://www.arabnews.com/node/1840096/business-economy.

Ashton, Kevin. *How to Fly a Horse: The Secret History of Creation, Invention, and Discovery*. New York: Doubleday, 2015.

Auerswald, Philip E., and Lewis M. Branscomb. "Valleys of Death and Darwinian Seas: Financing the Invention to Innovation Transition in the United States." *Journal of Technology Transfer* 28 (February 2003): 227–239. https://doi.org/10.1023/A:1024980525678.

Avent, Ryan. "Everything's Amazing and Nobody's Happy." *The Economist*, September 11, 2012. https://www.economist.com/free-exchange/2012/09/11/everything-amazing-and-nobodys-happy.

Badley, Graham. "A Place from Where to Speak: The University and Academic Freedom." *British Journal of Educational Studies* 57, no. 2 (June 2009): 146–163.

Bahcall, Safi. Loonshots: *How to Nurture the Crazy Ideas That Win*

Wars, Cure Diseases, and Transform Industries. New York: St. Martin's Press, 2019.

Banta, David. "What Is Technology Assessment?" *International Journal of Technology Assessment in Health Care* 25, no. S1 (July 2009): 7–9.

Bentham, Jeremy. *An Introduction to the Principles of Morals and Legislation*. Edited by J. H. Bums and H. L. A. Hart. London: Athlone Press, 1970.

Bloch, Ernst. *The Principle of Hope*. Cambridge, MA: MIT Press, 1995.

Bloch, Ernst. "Zur Ontologie des Noch-Nicht-Seins." *In Ernst Bloch—Auswahl aus seinen Schriften*, edited by Hans Heinz Holz, 63. Frankfurt a. M./Hamburg: Fischer, 1967.

Bloom, Nicholas, Charles I. Jones, John Van Reenen, and Michael Webb. "Are Ideas Getting Harder to Find?" *American Economic Review* 110, no. 4 (April 2020): 1104–1144.

Boissoneault, Lorraine. "Jurassic Park's Unlikely Symbioses with Real-World Science." *Smithsonian Magazine*, June 15, 2018. https://www.smithsonianmag.com/science-nature/jurassic-park-reveals-delicate-interplay-between-science-and-science-fiction-180969331/.

Bonvillian, William B. "The Connected Science Model for Innovation." In *The DARPA Model for Transformative Technologies*, edited by William B. Bonvillian, Richard Van Atta, and Patrick Windham, 77–144. Cambridge: Open Book Publishers, 2019.

Bonvillian, William B. "Power Play." *American Interest* 2, no. 2 (November 1, 2006). https://www.the-american-interest.com/2006/11/01/power-play/.

Bonvillian, William B., Richard Van Atta, and Patrick Windham, eds. *The DARPA Model for Transformative Technologies*. Cambridge: Open Book Publishers, 2019.

Borcherds, Richard E. "The Teapot Test for Quantum Computers." YouTube, 12:05, February 7, 2021. https://www.youtube.com/watch?v=sFhhQRxWTIM.

Boston University School of Medicine. "Framingham Heart Study." Accessed September 15, 2022. https://www.bumc.bu.edu/preventive-med/research/framingham-heart-study/.

Bowers, Simon. "Jeff Bezos: Amazon.com's 'Dread Pirate' Founder." *The Guardian*, August 6, 2013. https://www.theguardian.com/technology/2013/aug/06/jeff-bezos-amazon-washington-post.

Branscomb, Lewis M. *Investing in Innovation: Creating a Research and Innovation Policy That Works*. Cambridge, MA: MIT Press, 1998.

Bundesgesetzblatt. "Gesetz für den Ausbau erneuerbarer Energien (Erneuerbare-Energien-Gesetz-EEG 2014)." Bundesgesetzblatt Jahrgang 2014 Teil I Nr. 33. July 21, 2014.

Bundesministerium für Forschung und Bildung. "20 Milliarden für Bildung und Forschung." June 2, 2022. https://www.bmbf.de/bmbf/shareddocs/kurzmeldungen/de/2022/05/bmbf-haushalt-2022.html.

Carlyle, Thomas. *On Heroes, Hero-Worship, & the Heroic in History*. London: James Fraser, 1841.

"The Case for More State Spending on R&D." *The Economist*, January 16, 2021. https://www.economist.com/briefing/2021/01/16/the-case-for-more-state-spending-on-r-and-d.

CBInsights. "The Foie Gras'ing of Startups: Does Raising More VC Lead to Bigger Outcomes?" July 10, 2019. https://www.cbinsights.com/research/startup-vc-overfunding-foie-gras/.

"The Challenger." *The Economist*, March 15, 2018. https://www.economist.com/briefing/2018/03/15/the-challenger.

Chasan, Emily. "We Already Have the World's Most Efficient Carbon Capture Technology." Bloomberg. August 2, 2019. https://www.bloomberg.com/news/features/2019-08-02/we-already-have-the-world-s-most-efficient-carbon-capture-technology?leadSource=uverify%20wall.

Chesbrough, Henry: *Open Innovation: The New Imperative for Creating and Profiting from Technology*. PA, MA: Harvard Business Publishing, 2003.

Christensen, Clayton M. *The Innovator's Dilemma: When New Technologies Cause Great Firms to Fail*. Boston: Harvard Business Publishing, 1997.

Cixin, Liu. *The Three Body Problem*. New York: Tor Books, 2014.

Cohen, Jon. "Gentlemen of Science." *Science* 279, no. 5348 (January 9, 1998): 179. https://doi.org/10.1126/science.279.5348.179.

Collingridge, David. *The Social Control of Technology*. London: Pinter, 1982.

Collison, Patrick, and Tyler Cowen. "We Need a New Science of Progress." *The Atlantic*, July 30, 2019. https://www.theatlantic.com/science/archive/2019/07/we-need-new-science-progress/594946/.

Commission of Experts for Research and Innovation (EFI). *Research, Innovation and Technological Performance in Germany—EFI Report* 2022. Berlin: EFI, 2022.

ComputerScience.org. "Women in Computer Science: Getting Involved in STEM." Last modified September 9, 2022. https://www.computerscience.org/resources/women-in-computer-science/.

Cowen, Tylor. *The Great Stagnation: How America Ate All the Low-Hanging Fruit of Modern History, Got Sick, and Will (Eventually) Feel Better*. New York: Dutton Adult, 2011.

DARPA. "DARPA 2019 Strategic Framework." Defence Advanced Research Projects Agency. Accessed September 19, 2022. https://www.darpa.mil/attachments/DARPA-2019-framework.pdf.

"The Data Economy—Special Reports." *The Economist*, February 20, 2020. https://www.economist.com/special-report/2020-02-22.

Davis, Alex. "The Autonomous-Car Chaos of the 2004 Darpa Grand Challenge." *Wired*, January 6, 2021. https://www.wired.com/story/autonomous-car-chaos-2004-darpa-grand-challenge/.

Del Carpio, Lucia, and Maria Guadalupe. "More Women in Tech? Evidence from a Field Experiment Addressing Social Identity." IZA Institute of Labor Economics Discussion Paper Series, no. 11876 (October 2018). https://docs.iza.org/dp11876.pdf.

"A Dispute over COVID-Vaccine Technology Ends Up in Court." *The Economist*, September 1, 2022. https://www.economist.com/business/2022/09/01/a-dispute-over-covid-vaccine-technology-ends-up-in-court.

Drechsler, Wolfgang, Rainer Kattel, and Erik S. Reinert, eds. *Techno-Economic Paradigms: Essays in Honour of Carlota Perez*. London: Anthem Press, 2009.

Duckworth, Angela L. *Grit: The Power of Passion and Perseverance*. New York: Scribner, 2016.

Dugan, Regina E., and Kaigham J. Gabriel. "Changing the Business of Breakthroughs." *Issues in Science and Technology* 38, no. 4 (Summer 2022): 70–74.

Dugan, Regina E., and Kaigham J. Gabriel. "'Special Forces' Innovation: How DARPA Attacks Problems." *Harvard Business Review*, October 2013. https://hbr.org/2013/10/special-forces-innovation-how-darpa-attacks-problems.

Eberl, Ulrich. *Smart Machines: How Artificial Intelligence Is Changing Our Lives*. Munich: Hanser, 2016.

Edwards, Jane. "America's Frontier Fund to Invest in AI, Microelectronics to Secure US Tech Advantage." GovCon Wire, June 10, 2022. https://www.govconwire.com/2022/06/americas-frontier-fund-to-invest-in-ai-deep-tech-to-secure-us-tech-advantage/.

Endo, Akira. "A Gift from Nature: The Birth of the Statins." *Nature Medicine* 14 (October 2008): 1050–1052.

Endo, Akira. "A Historical Perspective on the Discovery of Statins." *Proceedings of the Japan Academy, Series B, Physical and Biological Sciences* 86, no. 5 (2010): 484–493. https://doi.org/10.2183/pjab.86.484.

Ericsson, Anders, and Robert Pool. *Peak: Secrets from the New Science of Expertise*. New York: HarperCollins, 2017.

Eschner, Kat. "This Wooden Running Machine Was Your Fixie's Great-Great Grandpa." *Smithsonian Magazine*, February 17, 2017. https://www.smithsonianmag.com/ smart-news/wooden-running-machine -was-your-fixies-great-great -grandpa-180962152/.

European Commission. "Data Act: Commission Proposes Measures for a Fair and Innovative Data Economy." February 23, 2022. https://ec.europa.eu/commission/ presscorner/detail/en/ip_22_1113.

European Commission. "Horizon Europe." Accessed September 17, 2022. https://research-and -innovation.ec.europa.eu/funding/ funding-opportunities/funding -programmes-and-open-calls/ horizon-europe_en.

Fallows, James. "The 50 Greatest Breakthroughs since the Wheel." *The Atlantic*, November, 2013. https://www.theatlantic.com/ magazine/archive/2013/11/ innovations-list/309536/.

Federal Agency for Disruptive Innovation SPRIND. "Die Binnen- windanlagen der Zukunft." Accessed September 19, 2022. https://www.sprind.org/de/projekte/ beventum/.

Federal Agency for Disruptive Innovation SPRIND. "Positions- papier IP-Transfer 3.0." July, 2022. https://www.sprind.org/cms/ uploads/220423_IP_Paper_ SPRIND_6f9b19a34f.pdf.

Federal Agency for Disruptive Innovation SPRIND. *SPRIND Podcast: Thomas Ramge spricht mit Menschen, die Neues neu denken."* Accessed September 18, 2022. https://www.sprind.org/de/podcast/.

Fernandez, Mar. Interview by Thomas Ramge, *SPRIND Podcast*, Federal Agency for Disruptive Innovation SPRIND. July 4, 2022. https://www.sprind.org/de/podcast/.

"A Firm Founded by Bill Gates Bets on a Novel Nuclear Reactor." *The Economist*, June 10, 2021. https://www.economist.com/science -and-technology/2021/06/10/a-firm -founded-by-bill-gates-bets-on -a-novel-nuclear-reactor.

Frantzman, Seth J. "Israel's Iron Dome Won't Last Forever." *Foreign Policy*, June 3, 2021. https:// foreignpolicy.com/2021/06/03/israels -iron-dome-wont-last-forever/.

Freeman, Christopher, John Clark, and Luc Soete. *Unemployment and Technical Innovation: A Study of Long Waves and Economic Develop- ment*. Westport, CT: Greenwood Press, 1982.

Free Software Foundation Europe. "Public Money? Public Code!" Accessed September 19, 2022. https://fsfe.org/activities/ publiccode/publiccode.en.html.

Füssel, Stefan. *Gutenberg*. London: Haus Publishing, 2019.

Gabriel, Ken. "ZEIT für Forschung: Sparking Innovation through Chal- lenges." ZEIT für Forschung. May

27, 2021. YouTube, 59:20. https://
www.youtube.com/
watch?v=ZJgrb1F1poE.

García, Jorge Luis, James J. Heck-
man, Duncan Ermini Leaf, and
María José Prados. "The Life-Cycle
Benefits of an Influential Early
Childhood Program." NBER Working
Paper 22993, December 2016.
http://www.nber.org/papers/w22993.

Gates, Bill. *How to Avoid Climate
Disaster*. New York: Knopf Doubleday, 2021.

Gentner, John. *The Idea Factory:
Bell Labs and the Great Age of
American Innovation*. New York:
Penguin, 2012.

Ghaffary, Shirin. "Big Tech Has a
Battle Ahead over Antitrust Regulation—and It's Going to Get
Messy." *Vox*, June 18, 2021.
https://www.vox.com/recode/
22537529/tech-battle-antitrust
-regulation-lina-khan-ftc-google
-facebook-apple-amazon-cicilline
-congress.

Gladwell, Malcolm. *Outliers. The
Story of Success*. New York:
Little, Brown, 2008.

Gnambs, Timo, and Barbara Hanf-
stingl. "The Decline of Academic
Motivation during Adolescence:
An Accelerated Longitudinal
Cohort Analysis on the Effect of
Psychological Need Satisfaction."
Educational Psychology 36, no. 9
(September 2016): 1698–1712.

Goldsmith, Barbara. *Obsessive
Genius: The Inner World of Marie
Curie*. New York: W. W. Norton,
2005.

Goldstein, Anna P., and Venkatesh
Narayanamurti. "Simultaneous
Pursuit of Discovery and Invention
in the US Department of Energy."
Research 47, no. 8 (October 2018):
1505–1512. https://doi.org/
10.1016/j.respol.2018.05.005.

Goldstein, Brett, and Lauren
Dyson, eds. *Beyond Transparency—
Open Data and the Future of Civic
Innovation*. San Francisco: Code
for America Press, 2013.

Gompers, Paul A. "The Rise and
Fall of Venture Capital." *Business
and Economic History* 23, no. 2
(Winter 1994): 1–26.

Gordon, Robert. *The Rise and Fall
of American Growth: The U.S.
Standard of Living Since the
Civil War*. Princeton, NJ: Princeton University Press, 2016.

Graham, Paul. "The Bus Ticket
Theory of Genius." November 2019.
http://www.paulgraham.com/genius
.html.

Grant, Adam. *Originals: How
Non-conformists Move the World*.
New York: Viking, 2016.

Grove, Andrew S. *Only the Paranoid
Survive: Lessons from the CEO of
INTEL Corporation*. New York:
Broadway Business, 1996.

Grunwald, Armin. *Technology Assessment in Practice and Theory*. Abingdon: Routledge, 2018.

Harari, Yuval N. *Sapiens: A Brief History of Humankind*. London: Harvill Secker, 2014.

Harhoff, Dietmar, et al. "Citation Frequency and the Value of Patented Inventions." *Review of Economics and Statistics* 81, no. 3 (August 1999): 511–515.

Henn, Steve. "When Women Stopped Coding." *NPR Morning Edition*, NPR, October 21, 2014. Audio, 4:33. https://www.npr.org/sections/money/2014/10/21/357629765/when-women-stopped-coding?t=1660301924425&t=1660302339887.

Holzki, Larissa. "Gebt Gründern Aufträge statt Fördermittel." *Handelsblatt*, March 17, 2021. https://www.handelsblatt.com/meinung/kommentare/kommentar-gebt-gruendern-auftraege-statt-foerdermittel/27013104.html#:~:text=Damit%20deutsche%20Start%2Dups%20Konzerne,wird%20auch%20das%20Finanzierungskapital%20folgen.

Hone, Robert, and John Palfreman. "Interview with Steve Jobs, 1990." GHB Archives, May 14, 1990. Video, 50:08. https://openvault.wgbh.org/catalog/V_AD9E0BC353BF435E83F28DEF165D4F40.

Isaacson, Walter. *The Innovators: How a Group of Hackers, Geniuses, and Geeks Created the Digital Revolution*. New York: Simon & Schuster, 2014.

Isaacson, Walter. *Steve Jobs*. New York: Simon & Schuster, 2011.

Israel, Paul. *Edison: A Life of Invention*. Hoboken, NJ: John Wiley & Sons, 1998.

Jacobs, Michael, and Mariana Mazzucato, eds. *Rethinking Capitalism: Economics and Policy for Sustainable and Inclusive Growth*. Hoboken, NJ: Wiley Blackwell, 2016.

Jarvis, Jeff. *Gutenberg the Geek*. Amazon, 2012. Kindle Edition.

Johnson, Steven. *How We Got to Now: Six Innovations That Made the Modern World*. New York: Riverhead Books, 2014.

Johnson, Steven. *Where Good Ideas Come From: The Natural History of Innovation*. New York: Riverhead Books, 2010.

Joint Research Centre. "China Overtakes the EU in High-Impact Publications." The European Commission's science and knowledge service. March 10, 2021. https://joint-research-centre.ec.europa.eu/system/files/2021-05/policy_brief_china_overtakes_the_eu_in_high_impact_publications_final_jrc124597.pdf

Jones, Elizabeth. "Sci-fi and Jurassic Park Have Driven Research, Scientists Say." The Conversation, June 10, 2015. https://theconversation.com/sci-fi-and-jurassic-park-have-driven-research-scientists-say-42864.

Kahneman, Daniel. *Thinking, Fast and Slow*. New York: Macmillan, 2011.

Kahneman, Daniel, and Amos Tversky. "Prospect Theory: An Analysis of Decision under Risk." *Econometrica* 47, no. 2 (1979): 263–291. https://doi.org/10.2307/1914185.

Kaplan, David A. *The Silicon Boys and Their Valleys of Dreams*. New York: William Morrow, 1999.

Karberg, Sascha. *Der Mann, der das Impfen neu erfand. Ingmar Hoerr, Curevac und der Kampf gegen die Pandemie*. Berlin: Aufbau, 2021.

Keynes, John Maynard. *The General Theory of Employment, Interest and Money*. London: Palgrave Macmillan, 1936.

Kimberly, John R., and Michael J. Evanisko. "Organizational Innovation: The Influence of Individual, Organizational, and Contextual Factors on Hospital Adoption of Technological and Administrative Innovations." *Academy of Management Journal* 24, no. 4 (December 1981): 689–713.

Kleinrock, Leonard. "The First Internet Connection with UCLA's Leonard Kleinrock." UCLA, January 13, 2022. YouTube, 07:42. https://www.youtube.com/watch?v=vuiBTJZfeo8&t=175s.

Kondratieff, Nikolai D. "The Long Waves in Economic Life." *Review of Economics and Statistics* 17, no. 6 (November 1935): 105–115.

Ksoll, Peter, and Vögtle, Fritz. *Marie Curie*. Hamburg: Rowohlt, 1988.

Kurz, Peter. *Weltgeschichte des Erfindungsschutzes: Erfinder und Patente im Spiegel der Zeiten*. Cologne: Carl Heymanns, 2000.

Leber, Rebecca. "The US Finally Has a Law to Tackle Climate Change." *Vox*, last modified August 16, 2022. https://www.vox.com/policy-and-politics/2022/7/28/23281757/whats-in-climate-bill-inflation-reduction-act.

Lee, Kai-Fu. *AI Superpowers: China, Silicon Valley, and the New World Order*. Boston, Houghton Mifflin Harcourt, 2018.

Le Pape, Amandine. "$8.5M to Accelerate Matrix, Polish Riot and Expand Modular." *Element* (blog), October 10, 2019. https://element.io/blog/8-5m-to-accelerate-matrix/.

Lepore, Jill. "The Disruption Machine: What the Gospel of Innovation Gets Wrong." *New Yorker*, June 23, 2014. https://www.newyorker.com/magazine/2014/06/23/the-disruption-machine.

Lewis, Michael. *The Undoing Project: A Friendship That Changed Our Minds*. New York: W. W. Norton, 2016.

Lotter, Wolf. *Innovation. Streit-schrift für barrierefreies Denken.* Hamburg: Körber, 2018.

Markoff, John. "Japanese Team Dominates Competition to Create Generation of Rescue Robots." *New York Times*, December 22, 2013. https://www.nytimes.com/2013/12/23/ science/japanese-team-dominates -competition-to-create-rescue -robots.html.

Marks, Paul. "Where Have All the Start-Ups Gone?" *Technologist*, April 3, 2017. https:// www.technologist.eu/where-have-all -the-start-ups-gone/.

Marquard, Odo. *Zukunft braucht Herkunft.* Stuttgart: Reclam, 2003.

Maslow, Abraham H. "A Theory of Human Motivation." *Psychological Review* 50, no. 4 (1943): 370–396.

Maslow, Abraham H., Henry Geiger, and Bertha G. Maslow. *The Farther Reaches of Human Nature.* New York: Compass, 1971.

Matich, Blake, and Jonathan Gifford. "Making Europe's Solar Future." *Green European Journal*, November 26, 2021. https://www .greeneuropeanjournal.eu/making -europes-solar-future/.

Mayer-Schönberger, Viktor, and Thomas Ramge. *Access Rules: Freeing Information to Stop Big Tech, Revive Innovation, and Emp-ower Society.* Berkeley: University of California Press, 2022.

Mayer-Schönberger, Viktor, and Thomas Ramge. "Are the Most Inno-vative Companies Just the Ones with the Most Data?" *Harvard Business Review*, February 7, 2018. https://hbr.org/2018/02/are-the -most-innovative-companies-just-the -ones-with-the-most-data.

Mayer-Schönberger, Viktor, and Thomas Ramge. "The Data Boom Is Here—It's Just Not Evenly Distri-buted." *MIT Sloan Management Review*, February 9, 2022. https:// sloanreview.mit.edu/article/ the-data-boom-is-here-its-just-not -evenly-distributed/.

Mayer-Schönberger, Viktor, and Thomas Ramge. *Reinventing Capita-lism in the Age of Big Data.* New York: Basic Books, 2018.

Mazzucato, Mariana. *The Entrepre-neurial State: Debunking Public vs. Private Myths.* London: Penguin Allen Lane, 2013.

Mazzucato, Mariana. *Mission Eco-nomy: A Moonshot Guide to Chan-ging Capitalism.* London: Penguin Allen Lane, 2021.

McCraw, Thomas K. *Prophet of Innovation.* Cambridge, MA: Harvard University Press, 2007.

McMahon, Liv. "Ministry of Defense Acquires Government's First Quan-tum Computer." BBC, June 9, 2022. https://www.bbc.com/news/technology -61647134.

Microsoft. "Microsoft Acquires GitHub." Microsoft News Center,

June 4, 2018. https://news.microsoft
.com/2018/06/04/microsoft-to
-acquire-github-for-7-5-billion/.

Mirchandani, Vinnie. *The New
Polymath: Profiles in Compound-
Technology Innovations*. Hoboken,
NJ: John Wiley & Sons, 2010.

National Library of Medicine
(NLM). "Dr. Alan Kay Talks about
the Future at Annual Lindberg-
King Lecture." NLM In Focus,
October 4, 2018. https://infocus
.nlm.nih.gov/2018/10/04/dr-alan-kay
-talks-about-the-future-at-annual
-lindberg-king-lecture/.

National Science Foundation.
"Science and Engineering Degrees:
1966–2010." National Center for
Science and Engineering Statistics. Last modified June 2013.
https://www.nsf.gov/statistics/
nsf13327/content.cfm?pub_
id=4266&id=2.

Nicholas, Tom. *VC: An American
History*. Cambridge, MA: Harvard
University Press, 2019.

Niedermeyer, Edward. *Ludicrous:
The Unvarnished Story of Tesla
Motors*. Dallas: BenBella Books,
2019.

OECD. "Gross Domestic Spending on
R&D." Accessed September 17, 2022.
https://data.oecd.org/rd/gross
-domestic-spending-on-r-d.htm.

OECD. "Public Procurement for
Innovation: Good Practices and
Strategies." OECD Public Gover-
nance Reviews, June 2, 2017.

https://www.oecd.org/gov/public
-procurement-for-innovation
-9789264265820-en.htm.

Official Journal of the European
Union. "Consolidated Version of
the Treaty on the Functioning of
the European Union." Section 2,
Art. 107, No. 1 (October 10, 2012).

O'Mara, Margaret. *The Code: Silicon Valley and the Remaking of
America*. New York: Penguin Press,
2019.

O'Mara, Margaret. "Why Can't Tech
Fix Its Gender Problem?" *MIT Technology Review*, August 11, 2022.
https://www.technologyreview
.com/2022/08/11/1056917/tech-fix
-gender-problem/.

Overly, Steven. "This Government
Loan Program Helped Tesla at a
Critical Time: Trump Wants to Cut
It." *Washington Post*, March 16,
2017. https://www.washingtonpost.
com/news/innovations/wp/2017/03/16/
this-government-loan-program-helped
-tesla-at-a-critical-time-trump
-wants-to-cut-it/.

Owens, Brian. "Canada Announces
New Innovation Agency—and It's Not
Modelled on DARPA." *Nature*, April
27, 2022. https://www.nature.com/
articles/d41586-022-01190-4.

Pain, Elizabeth. "French Science
Bill Promises Boost to Public
R&D." *Science*, July 24, 2020.

https://www.science.org/content/
article/french-science-bill
-promises-boost-public-rd.

Paul, Kari. "Elizabeth Holmes Trial: Jury Finds Theranos Founder Guilty on Four Fraud Counts." *The Guardian*, January 4, 2022.

https://www.theguardian.com/technology/2022/jan/03/elizabeth-holmes-trial-jury-finds-theranos-founder-guilty-on-four-counts.

Peneder, Michael, and Andreas Resch. "Schumpeter and Venture Finance: Radical Theorist, Broke Investor, and Enigmatic Teacher." *Industrial and Corporate Change* 24, no. 6 (December 2015): 1315–1352. https://doi.org/10.1093/icc/dtv004.

Perez, Caroline C. *Invisible Women: Exposing Data Bias in a World Designed for Men*. New York: Vintage, 2020.

Philippon, Thomas. *The Great Reversal: How America Gave Up on Free Markets*. Cambridge, MA: Harvard University Press, 2019.

Portincaso, Massimo, Antoine Gourévitch, Arnaud de la Tour, Arnaud Legris, Thomas Salzgeber, and Tawfik Hammoud. "The Deep Tech Investment Paradox: A Call to Redesign the Investor Model." Boston Consulting Group and Hello Tomorrow, May 2021. https://hello-tomorrow.org/wp-content/uploads/2021/05/Deep-Tech-Investment-Paradox-BCG.pdf.

Pratt, Gill. "DARPA Challenge History." Humanoid Robot, January 1, 2014. YouTube, 3:09. https://www.youtube.com/watch?v=MGMk4SdLGjE.

PRC State Council. "China's Spending on R&D Reaches New High in 2021." The People's Republic of China (PRC) State Council Statistics. Last modified January 26, 2022. http://english.www.gov.cn/archive/statistics/202201/26/content_WS61f0de8cc6d09c94e48a44d1.html#:~:text=The%20country's%20total%20expenditure%20on,%2Dyear%2C%20said%20the%20NBS.

PRC State Council. "Notice of the State Council on the Publication of 'Made in China 2025.'" CSET Center for Security and Emerging Technology. May 8, 2015. https://cset.georgetown.edu/wp-content/uploads/t0432_made_in_china_2025_EN.pdf.

Prillaman, McKenzie. "Billions More for US Science: How the Landmark Spending Plan Will Boost Research." *Nature* 608, no. 7922 (August 2022): 249.

Ramge, Thomas. *Augmented Intelligence: Wie wir mit Daten und KI besser entscheiden*. Stuttgart: Reclam, 2020.

Ramge, Thomas. "Das Innovations-Paradox." *brand eins*, December 2020.

Ramge, Thomas. "Gestopft!" *brand eins*, May 2021.

Ramge, Thomas. "Not Invented Here: Frank Piller im Interview über Open Innovation." *brand eins*, January 2017.

Ramge, Thomas. *Postdigital: Using AI to Fight Coronavirus, Foster Wealth and Fuel Democracy.* Hamburg: Murmann, 2020.

Ramge, Thomas. *Who's Afraid of AI? Fear and Promise in the Age of Thinking Machines.* New York: The Experiment, 2019.

Ramge, Thomas. "Wie nützt das Neue?" *brand eins*, July 2013. https://www.brandeins.de/magazine/brand-eins-wirtschaftsmagazin/2013/fortschritt-wagen/wie-nuetzt-das-neue.

Reuters. "France Recruits Dassault Systemes, OVH for Alternative to U.S. Cloud Firms." October 3, 2019. https://www.reuters.com/article/us-france-dataprotection-idUSKBN1WI189.

Ridley, Matt. *How Innovation Works: And Why It Flourishes in Freedom.* New York: HarperCollins, 2020.

Ridley, Matt. *The Rational Optimist: How Prosperity Evolves.* New York: HarperCollins, 2010.

Robertson, Adi. "How the EU Is Fighting Tech Giants with Margrethe Vestager." *The Verge*, March 17, 2022. https://www.theverge.com/22981261/margrethe-vestager-decoder-antitrust-eu-apple-facebook-google-jedi-blue.

Rothman, Joshuah. "Are Things Getting Better or Worse?" *New Yorker*, July 23, 2018. https://www.newyorker.com/magazine/2018/07/23/are-things-getting-better-or-worse.

S&P Global. "Top 250–Global Energy Company Rankings." Accessed September 20, 2022. https://www.spglobal.com/commodityinsights/top250/rankings.

Sattelberger, Thomas. "Warum Deutschlands Forschungsinstitute 'fette Katzen' sind." *Manager Magazin*, November 22, 2018. https://www.manager-magazin.de/unternehmen/max-plack-fraunhofer-helmholtz-mangelnde-effizienz-in-der-forschung-a-00000000-0002-0001-0000-000160904402.

Saxenian, AnnaLee. *Regional Advantage. Culture and Competition in Silicon Valley and Route 128.* Cambridge, MA: Harvard University Press, 1994.

Schäfer, Annette. *Die Kraft der schöpferischen Zerstörung. Joseph A. Schumpeter. Die Biografie.* Frankfurt a.M.: Campus, 2008.

Schäfer, Rene. "Das Kasino ist vorübergehend geschlossen." Interview by Thomas Ramge. *brand eins*, July 2022. https://www.brandeins.de/magazine/brand-eins-wirtschaftsmagazin/2022/wendepunkte/start-ups-in-der-krise-das-kasino-ist-voruebergehend-geschlossen.

Schlagenhauf, Karl, and Fanji Gu. *The Brain and AI: Correspondence between a German Engineer and a Chinese Scientist.* Shanghai: Shanghai Educational Publishing, 2019.

Schmidt, Eric, and Jonathan Rosenberg. *How Google Works.* New York: Grand Central Publishing, 2014.

Schumpeter, Joseph A. *Business Cycles*. New York: McGraw-Hill, 1939.

Schumpeter, Joseph A. *Capitalism, Socialism and Democracy*. New York: HarperCollins, 1942.

Schumpeter, Joseph A. *History of Economic Analysis*. Abingdon: Routledge, 1954.

Schwochow, Jan, and Thomas Ramge. *The Global Economy as You've Never Seen It*. New York: The Experiment, 2018.

Shapiro, Carl, and Hal Varian. *Information Rules: A Strategic Guide to the Network Economy*. Boston: Harvard Business Publishing, 1998.

Smith, Noah. "Interview with Patrick Collison, Co-founder and CEO of Stripe." Noahpinion, May 8, 2020. https://noahpinion.substack.com/p/interview-patrick-collison-co-founder.

Squatriglia, Chuck. "Feds Lend Tesla $465 Million to Build Electric Car." *Wired*, June 24, 2009. https://www.wired.com/2009/06/tesla-loan/.

Staab, Philipp. *Digitaler Kapitalismus: Markt und Herrschaft in der Ökonomie der Unknappheit*. Berlin: Suhrkamp, 2019.

Stephenson, Neal. *Snow Crash*. New York: Bantam Books, 1992.
Stokes, Donald E. *Pasteur's Quadrant–Basic Science and Technological Innovation*. Washington, DC: Brookings Institution Press, 1997.

Swiss Federal Council. "Bundesverwaltung beschafft maschinelles Übersetzungsprogramm." December 18, 2019. https://www.admin.ch/gov/de/start/dokumentation/medienmitteilungen.msg-id-77610.html.

Syed, Matthew. *Rebel Ideas: The Power of Diverse Thinking*. London: John Murray, 2019.

Synopsys. "2020 Open Source Security and Risk Analysis Report." Accessed September 19, 2022. https://ttpsc.com/wp3/wp-content/uploads/2020/10/2020-ossra-report.pdf.

Tabarrok, Alex. *Launching the Innovation Renaissance: A New Path to Bring Smart Ideas to Market Fast*. New York: TED, 2011.

Taleb, Nasim N. *The Black Swan: The Impact of the Highly Improbable*. New York: Random House, 2007.

Teachout, Zephyr. *Break 'Em Up: Recovering Our Freedom from Big Ag, Big Tech, and Big Money*. New York: Macmillan, 2020.

Teare, Gené. "VC Funding to Black-Founded Startups Slows Dramatically as Venture Investors Pull Back." Crunchbase, June 17, 2022. https://news.crunchbase.com/diversity/vc-funding-black-founded-startups/.

"Technology Firms Are Both the Friend and the Foe of Competition." *The Economist*, November 15, 2018. https://www.economist.com/special-report/2018/11/15/technology-firms-are-both-the-friend-and-the-foe-of-competition.

Thalidomide Society, The. "About Thalidomide." Accessed September 21, 2022. https://thalidomidesociety .org/what-is-thalidomide/.

Theiner, Peter. *Robert Bosch: Unternehmer im Zeitalter der Extreme*. Munich: C. H. Beck, 2019.

Thiel, Peter, and Blake Masters. *Zero to One*. New York: Random House, 2014.

"TikTok and the Sino-American Tech Split." *The Economist*, July 9, 2020. https://www.economist.com/ leaders/2020/07/09/tiktok-and -the-sino-american-tech-split.

Toffler, Alvin, and Adelaide Toffler. *Future Shock*. New York: Random House, 1970.

Tollefson, Jeff. "The Rise of 'ARPA-Everything' and What It Means for Science." *Nature* 595, no. 7868 (July 2021): 483–484.

Trbuhović-Gjurić, Desanka. *Im Schatten Albert Einsteins: Das tragische Leben der Mileva Einstein-Marić*. Bern/Stuttgart/ Wien: Paul Haupt, 1993.

Tse, Edward. *China's Disruptors: How Alibaba, Xiaomi, Tencent, and Other Companies Are Changing the Rules of Business*. New York: Random House, 2016.

Ulmann, Bernd. Interview by Thomas Ramge, *SPRIND Podcast*, Federal Agency for Disruptive Innovation SPRIND. April 26, 2021. Audio, 50:44. https://www.sprind .org/de/podcast/.

UNESCO Institute for Statistics. "Research and Development Expenditure (% of GDP)." World Bank. Accessed September 18, 2022. https://data.worldbank.org/ indicator/GB.XPD.RSDV.GD.ZS.

United Nations. "Transforming Our World: The 2030 Agenda for Sustainable Development." Accessed September 18, 2022. https:// sdgs.un.org/2030agenda.

US Bureau of Labor Statistics. "Labor Force Statistics from the Current Population Survey." Accessed September 16, 2022. https:// www.bls.gov/cps/cpsaat11.htm.

Vaish, Esha, and Simon Jessop. "Spotify Prompts Nordic Pension Funds to Add Private Equity to Playlists." Reuters, April 15, 2019. https://jp.reuters.com/ article/us-nordics-privateequity -pensions-analys-idUSKCN1RR1QM.

Vance, Ashlee. *Elon Musk: How the Billionaire CEO of SpaceX and Tesla Is Shaping Our Future*. New York: HarperCollins, 2015.

Verniero, Madeline. "Does DARPA's Legal Structure Lead to Ethical Lapses?" *The Regulatory Review*, June 16, 2022. https://www .theregreview.org/2022/06/16/ verniero-does-darpas-legal-structure -lead-to-ethical-lapses/.

Vinsel, Lee, and Andrew Russell. *The Innovation Delusion: How Our Obsession with the New Has Disrupted the Work That Matters Most*. New York: Crown Publishing, 2020.

Wakabayashi, Daisuke, and Scott Shane. "Google Will Not Renew Pentagon Contract That Upset Employees." *New York Times*, June 1, 2018. https://www.nytimes.com/2018/06/01/technology/google-pentagon-project-maven.html.

Warnke, Philine, Kerstin Cuhls, Ulrich Schmoch, Lea Daniel, Liviu Andreescu, Bianca Dragomir, and Radu Gheorghiu. "100 Radical Innovation Breakthroughs for the Future: The Radical Innovation Breakthrough Inquirer." European Commission, June 24, 2019. https://euraxess.ec.europa.eu/worldwide/india/ec-publishes-forsesight-study-100-radical-innovation-breakthroughs-future.

Wattendorf, Dan. Interview by Thomas Ramge, *SPRIND Podcast*, Federal Agency for Disruptive Innovation SPRIND. June 7, 2021. Audio, 45:18. https://www.sprind.org/de/podcast/.

Watzinger, Martin, Thomas A. Facker, Markus Nagler, and Monika Schnitzer. "How Antitrust Enforcement Can Spur Innovation: Bell Labs and the 1956 Consent Decree." *American Economic Journal* 12, no. 4 (November 2020): 328–359.

Weber, Max. *Economy and Society*. Cambridge, MA: Harvard University Press, 2019.

Weiser, Mark. "The Computer for the 21st Century." *Scientific American* 265, no. 3 (September 1991): 94–104.

Welzer, Harald. "Hinterm Horizont." *SZ Magazin*, October 25, 2015. https://sz-magazin.sueddeutsche.de/leben-und-gesellschaft/hinterm-horizont-81799.

White House, the. "The Biden-Harris Administration FY 2023 Budget Makes Historic Investments in Science and Technology." White House Press Releases, April 5, 2022. https://www.whitehouse.gov/ostp/news-updates/2022/04/05/the-biden-harris-administration-fy-2023-budget-makes-historic-investments-in-science-and-technology/.

WHO. *Global Status Report on Road Safety 2018*. Geneva: World Health Organization, 2018.

Widdicombe, Lizzie. "The Rise and Fall of WeWork," *New Yorker*, November 6, 2019. https://www.newyorker.com/culture/culture-desk/the-rise-and-fall-of-wework.

Wilson, James Q. *Bureaucracy: What Government Agencies Do and Why They Do It*. New York: Basic Books, 1991.

Wooldridge, Adrian. "The Coming Tech-Lash." *The Economist*, November 18, 2013. https://www.economist.com/news/2013/11/18/the-coming-tech-lash.

Zak, Paul. "Measurement Myopia." Drucker Institute, April 7, 2013. https://www.drucker.institute/thedx/measurement-myopia/.

Page 14 (above) Tab. XXI c.
Die Buchdruckerei, circa 1770,
Daniel Chodowieki: 62 bisher
*unveröffentlichte Handzeichnungen
zu dem Elementarwerk von Johann
Bernhard Basedow.* Voigtländer-
Tetzner, Frankfurt am Main, 1922.

Page 14 (below) Hans Linde.
https://pixabay.com/de/photos/
mähdrescher-ernte-korn-4401822/.

Page 110/111 Solvay Conference,
1927 (topic: newly developed
quantum theory) http://w3.pppl.gov/
http://doi.org/10.3932/ethza
-000046848, Benjamin Couprie,
Institut International de Physique
de Solvay.

Page 134 (above) The French
biologist Louis Pasteur
(1822–1895), 1878 (detail).
http://history.amedd.army.mil/
booksdocs/misc/evprev.

Page 134 (left) Niels Bohr
(1885–1962), Danish physicist.
© The Nobel Foundation, Stockholm.
AB Lagrelius & Westphal.

Page 134 (below) The inventor
Thomas Alva Edison (1847‑1931),
circa 1922.
Louis Bachrach, Bachrach Studios,
restored by Michel Vuijlsteke.

Page 236/237 Pexels:
"Planet Earth Close Up Photo"
is licensed under CC0.

Page 282 (left) Laguna de la Vera.
© SPRIND GmbH (right) Thomas Ramge.
© Peter van Heesen.

ACKNOWLEDGMENTS

For long discussions, smart ideas, reading suggestions, research support, and constructive critical feedback, we would like to warmly thank Markus Albers, Martin Chaumet, Erich Clementi, Jano Costard, Kerstin Cuhls, Roland Damann, Barbara Diehl, Regina Dugan, Christian Egle, Mar Fernández, Kaigham (Ken) Gabriel, Chris Gernreich, Dietmar Harhoff, Justus Haucap, Katharina Hölzle, Ingmar Hoerr, Elke Jensen, Christoph Koch, Lili and Paco Laguna, Viktor Mayer-Schönberger, Rebecca Saive, René Schäfer, Karl Schlagenhauf, Jelka Seitz, Bernd Ulmann, Dan Wattendorf, and Yvonne Winter. Thanks to Alexander Voss for his tireless help with research and fact checking. We would like to thank Mike Meiré, Tim Giesen, and Julia Pidun for their creativity and implementation of the design, typesetting, and infographics. We are thankful to Nobel Laureate Stefan Hell for contributing his foreword. Gita Manaktala and David Weinberger deserve great thanks for the trust and freedom that we had with this book project.

Rafael Laguna de la Vera
is the founding director
of the Federal Agency
for Disruptive Innovation
(SPRIN-D). At age six-
teen, he founded his
first start-up, Elephant
Software. He built up
numerous other technology
companies and has worked
as a technology investor,
interim manager, and
consultant for venture
capital funds. His
involvement with Open-
Xchange AG and SUSE Linux
established his reputa-
tion as an open source
pioneer and champion of
an open internet. Laguna
is a visiting professor
at several universities
and cofounding partner
of CODE University of
Applied Sciences.

Thomas Ramge has authored
twenty books on inno-
vation, technology, and
transformation.
His essays and longreads
are being published
in MIT Sloan Management
Review, The Economist,
Harvard Business Review,
Foreign Affairs, brand
eins, Die Zeit, Frank-
furter Allgemeine Zeitung,
Die Welt, among others.
He has received numerous
book and journalism
awards, including the
Best Business Book Award
on Innovation and Tech-
nology, the Axiom Business
Book Award, the getAbstract
International Book Award,
the German Business Book
Award, and the Herbert
Quandt Media Award. Ramge
is an associated member
of the Einstein Center
for Digital Future and an
Alumni Senior Research
Fellow at the Weizenbaum
Institute for the Net-
worked Society. He lives
in Berlin with his wife
and son.